MAKING PODCASTS

Essential Library
An Imprint of Abdo Publishing
abdobooks.com

BY SUE BRADFORD EDWARDS

abdobooks.com

Printed in the United States of America, North Mankato, Minnesota.
052024
092024

THIS BOOK CONTAINS RECYCLED MATERIALS

Cover Photos: Shutterstock Images (microphone, phone, headphones, soundwaves)
Interior Photos: Shutterstock Images, 1 (headphones), 1 (soundwaves), 19, 38, 57, 88; Phynart Studio/iStockphoto, 4; Seemanta Dutta/Alamy, 6–7; Wayhome Studio/Shutterstock Images, 9; South Agency/E+/Getty Images, 11; Jade Gao/AFP/Getty Images, 12–13; Roy Morsch/The Image Bank/Getty Images, 16; John Green/Bay Area News/MediaNews Group/Getty Images, 22–23; Gabby Jones/Bloomberg/Getty Images, 25; Suzanne Kreiter/Boston Globe/Getty Images, 27; Vivien Killilea/Getty Images for SiriusXM/Getty Images Entertainment/Getty Images, 30; Kit Leong/Shutterstock Images, 34–35; Jakub Porzycki/NurPhoto/Getty Images, 37; Scott Strazzante/San Francisco Chronicle/AP Images, 41; Hyoung Chang/MediaNews Group/Denver Post/Getty Images, 42; Sheldon Cooper/SOPA Images/LightRocket/Getty Images, 45; Brad Barket/Getty Images for Vulture Festival/Getty Images Entertainment/Getty Images, 48–49; Nejron Photo/Shutterstock Images, 51; Patti McConville/Alamy, 52; Jerry Larson/Waco Tribune Herald/AP Images, 54; Liz Hafalia/San Francisco Chronicle/Hearst Newspapers/Getty Images, 59; John B Hewitt/Alamy, 60; Nuccio DiNuzzo/Chicago Tribune/Tribune News Service/Getty Images, 64; Jonas Walzberg/dpa/picture alliance/Getty Images, 67; Chris Pizzello/Invision/AP Images, 70–71; Iain Masterton/Alamy, 75; Gerry Hanan/Getty Images for SXSW/Getty Images Entertainment/Getty Images, 76; Dragon Images/Shutterstock Images, 79; sjscreens/Alamy, 82–83; Jeff Gritchen/Orange County Register/MediaNews Group/Getty Images, 85; Joe Purchase/Alamy, 90; Julia Weeks/AP Images, 92; Alisha Jucevic/The Washington Post/Getty Images, 94–95; Gina Kelly/Alamy, 100; Kaspars Grinvalds/Shutterstock Images, 101

Editor: Laura Stickney
Series Designer: Maggie Villaume

Library of Congress Control Number: 2023949559

Publisher's Cataloging-in-Publication Data
Names: Edwards, Sue Bradford, author.
Title: Making podcasts / by Sue Bradford Edwards
Description: Minneapolis, Minnesota: Abdo Publishing, 2025 | Series: Making media | Includes online resources and index.
Identifiers: ISBN 9781098293376 (lib. bdg.) | ISBN 9798384912644 (ebook)
Subjects: LCSH: Podcasts--Juvenile literature. | Podcasting--Juvenile literature. | User-generated content--Juvenile literature. | Web publishing--Juvenile literature. | Streaming technology (Telecommunications)--Juvenile literature.
Classification: DDC 791.46--dc23

CONTENTS

ON AIR

CHAPTER **ONE**

PODCAST PROJECT

High school students in the advanced media class moved around the room, finding seats next to their partners. As they regrouped, the students chatted about their podcast projects. It took a few minutes before everyone in the class was settled.

Reba was paired with Ari for the project. She asked him if he had any ideas about potential podcast topics. Ari explained what he'd heard other students discussing. Bryan and Mark were making a podcast about high school sports. They planned to interview athletes, including alumni who had gone on to play at the college level. Melanie and Nat's podcast would focus on science fair projects, and they planned to talk to teachers about what judges look for in a winning project.

Reba didn't know what topic she wanted to podcast about. So she and Ari started brainstorming. Ari said that his mother had suggested several different topics, including politics and global warming. These were the types of podcasts his mother usually listened to when she was jogging.

Today, most podcast listeners are young people. Some schools use podcasts as learning tools, such as offering podcast versions of lessons and teaching students how to create their own podcasts.

Reba shook her head. "Those are good ideas," she said. "But I don't think they're right for this project. I wish someone would ask me questions to get the ideas flowing."

As soon as Reba mentioned asking questions, Ari got an idea. He suggested that they do an interview-based podcast. Ari and Reba would talk to guests about questions they wished people would ask them. This would be a great way to let interviewees share their views about issues they rarely got to talk about. They could call the podcast *I Wish Someone Would Ask*. They could talk to their fellow students and other people who they knew.

Reba quickly wrote down their ideas, and the pair continued to brainstorm. Soon they had a list of 14 people to interview for their podcast. According to the assignment guidelines, Reba and Ari needed to make a sample episode and upload it to the class website. They decided to go to Reba's house after school to record the episode.

Some podcasters use voice recording apps on their phones, such as the Voice Memos app that comes preinstalled on the iPhone.

RECORDING THE PODCAST

At Reba's house, the pair sat down at the kitchen table. Reba unlocked her phone and opened the Voice Recorder app. She set her phone down on the table and hit the "record" button. Then she looked up at Ari. "Tell me," Reba said. "What is it you wish people would ask you?"

Ari stuttered and stammered. He gulped down a swallow of water and then managed to speak. "People always want to know what I want to be," he said. "They act like it's off in the future just because I'm 16. But I wish people would ask me about what I am now."

"What else?" prompted Reba.

"What do you mean what else?" asked Ari. "That's it."

Reba paused the recording. Then she looked at the timer. She told Ari that the interview had lasted less than a minute, which was a problem. They needed to record a ten-minute sample for their assignment.

Ari said Reba should have warned him that they were about to begin recording. He also suggested it would be a good idea to

THIS TEENAGE LIFE

Today, some teens record their own podcasts. *This Teenage Life* was originally made by a group of students from High Tech High in San Diego, California, working with teacher Molly Josephs. The first group of teens involved in the podcast has aged out of the project and moved on to other things. Teens throughout the United States, India, and other countries have continued the project. Periodically, the podcast's blog invites listeners to send an email if they want to participate in a discussion. The podcast works to create connections so that teens don't feel isolated. In 2023, the podcast was on season five. It had covered a variety of topics, such as social anxiety and college applications.

come up with some questions for Reba to ask. This would help them stay on topic during the podcast recording.

Together, Reba and Ari came up with a list of questions. The list included questions such as “What is it you wish people would ask you?” and “Why is it important to you?” They worked hard to develop questions that would apply to a wide variety of people. Then Ari suggested that they listen to the recording they had already made.

Reba pressed the “play” button on her phone. As they listened to the recording, the two were surprised at how quiet their voices sounded. There was also a loud humming noise in the background. “I think that’s the refrigerator running,” said Reba. “Maybe we should record this in the living room.”

They repeated the process, recording Reba asking the first question and Ari responding. Then they listened to their sample again. This time, their voices sounded a little louder because they had spoken up. But the bubbling of the aquarium could be heard in the background of the recording. “I don’t think this is good enough,” said Reba.

PODCRAFT

Given the variety of topics that podcasts cover, it probably isn’t surprising that there are several podcasts about podcasting. One of them is called *PodCraft*. In 2023 it was in its seventeenth season. The podcast covers all things podcasting with host Colin Gray, who is also the founder of the *Podcast Host*. This is a well-established blog about podcasting. Some *PodCraft* episodes feature listener questions, time-saving tips, and information about how to work more efficiently. In addition to podcasts, the site includes articles about how podcasting has changed and the necessary gear to get started.

Many podcasters use adjustable stands to keep their microphones in place. Some also use a pop filter, a type of screen that helps soften "popping" sounds like "p" or "b."

Ari looked up a video about podcasting on his phone to see what they could do to improve the sound of their recording. "This video says we should be using a microphone," he told Reba.

Neither one of them had a microphone, so Reba texted her uncle Vince. He worked at the local university in the media communications department. He oversaw several different podcasts that the university produced, including one about getting into college and one called *The Secrets in College Halls,* which explored surprising things that professors study.

Vince agreed to come over to the house to help Reba and Ari with their project. When Vince arrived, he pulled his laptop and a microphone out of his backpack. He explained that with these two pieces of equipment, Reba and Ari would have everything they needed to record and edit their podcast.

Reba and Ari opened an audio recording and editing program called Audacity on the laptop. They started a new recording and spoke into the microphone, going through all the questions on their list. This time, when they listened to the recording, their voices sounded clear and there were no background noises. But the recording program showed that they had recorded only nine minutes of audio, and they needed ten minutes.

Vince explained that they could add an intro to the beginning of the recording. Then they could add an outro to the end. In the outro, they would thank their listeners

Some podcasters wear headphones while recording. This helps them manage their voice levels and block out outside noises.

6:19.430
MacBook Pro

and anyone who helped them with the podcast. After a quick discussion, Ari recorded the intro and Reba recorded the outro. Ari gave Reba a thumbs-up when she thanked Vince, giving him credit as their sound engineer.

EDITING THE PODCAST

Now Reba and Ari were ready to edit their podcast. Vince warned them to make copies of all their files instead of editing the original files. He made sure they understood that Audacity does what is called destructive editing. This means that if a user cuts out a portion of audio and then saves the file, there is no way to get the cut audio back. Vince explained that this was why he never edited an original file. He always worked only with copies.

Vince sat back and let Reba and Ari begin the work of editing their podcast. First, they added the intro audio to their recording. Then they watched a tutorial video about using transitions and effects. Once Reba and Ari understood how some of the controls worked, they created an effect that

When editing a podcast, a creator may need to consider deleting parts of the audio recording, such as awkward pauses or background noises. The creator may also need to add sound effects or adjust the volume.

PODCAST EDITING

Not everyone enjoys editing their podcasts. That is why many professional podcasters pay someone else to do this job. Some podcasters record their podcasts in order, starting with the intro, moving on to the episode itself, and then doing the outro. They may use software such as Audacity to minimize background noise in their recording, but they don't rearrange parts of the recording or insert one recording into another. There are no added sound effects. Audio producer Matthew McLean advises podcasters who dislike editing to do as little of it as possible. He warns that if these podcasters try to edit too much, it could cause them to procrastinate or feel stressed.

allowed them to fade into the beginning of the podcast, with the volume starting out quiet and then growing louder.

Then they added the outro and faded out to silence at the very end. Ari did a quick Google search for sound effects and found what he wanted for free on a website called Pixabay. He downloaded the sound effect of a pen scratching on paper. Then he found the point in the podcast where he talked about writing poetry. He inserted the sound effect into their audio file.

Reba briefly took over the computer and found another sound effect. This one was the sound of a clock ticking. She spliced it in right before the outro, in which she said it was time to go and thanked everyone for listening.

Reba and Ari had been working on their podcast most of the evening, and they were happy with the results. They knew it wasn't as good as a professional podcast, but they thought it was a great first attempt. They uploaded the podcast to the class website. Several other podcasts had already been uploaded, so Reba and Ari clicked on them to hear what their classmates had created.

The sound quality in one of the podcasts wasn't very good, so the pair messaged their classmates asking if they had used a microphone. Next, Reba and Ari tuned in to Melanie and Nat's science fair podcast. The sound quality was excellent, and they had included special effects that sounded like the zapping, crackling electricity noise in a Frankenstein movie. One podcast was about local politics at the school board level. Two people running for the board were debating book banning. Another group had created a serialized fiction podcast. Each episode would be a chapter of a story they were coauthoring. In yet another podcast, people reviewed a local restaurant, pointing out dishes that were affordable for teens who might be paying for their own meal.

> **"It's never been easier to get into editing your own audio.[1]"**
>
> ***—Mike Russell of audio company Music Radio Creative, 2022***

Reba and Ari looked forward to hearing everything that their classmates created. They also couldn't wait to share their own podcast with everyone. They were impressed by the variety of podcasts that could be made with some simple equipment and a creative idea.

THE BLACK CROWES

CHAPTER TWO

THE HISTORY OF PODCASTS

For podcasting to become possible, people needed an easy way to access and play recordings on the go. In 1979, the electronics company Sony invented the Walkman. This portable cassette player allowed people to carry their music around with them. The first Walkman was made of aluminum and was only slightly larger than the cassette tapes it played. The Walkman had no speaker, so Sony included headphones that were small enough for someone to wear while walking around. Later Walkman products played compact discs (CDs). Sony sold 400 million Walkmans, including 200 million of the early cassette versions.[1]

Portable music players allowed people to listen to recordings on the go. But they still needed to use physical tapes and CDs. In the 1980s, researchers laid the groundwork for the digital music files that would eventually make podcasts possible. In 1983, Karlheinz Brandenburg was a student at Friedrich-Alexander University in Erlangen-Nuremberg, Germany. He was working on a PhD in electrical engineering. His adviser challenged him to study how to transmit music digitally

The first Sony Walkman was called the Walkman TPS-L2. In June 1980, it was released in the United States, where it quickly became a hit. Throughout the 1980s, Sony made several different Walkman products.

along telephone lines. To do so, they needed to find a file format that preserved the depth and quality of the music. Brandenburg worked on this project after he graduated, teaming up with a variety of companies, including AT&T Bell Laboratories.

It wasn't until 1992 that the team perfected MP3 files. These small files save space by deleting sounds beyond human hearing and compressing any remaining frequencies. Before MP3 files, sound recordings captured sounds that were too high or low in pitch for people to hear.

In 1998, two companies released MP3 players. South Korean company Saehan Information Systems made the MPMan F10. Diamond Multimedia made the Rio PMP300. Neither player became popular because it took hours to load MP3 files onto them.

In February 2001, Apple hired engineer Tony Fadell to help design an MP3 player. Apple CEO, cofounder, and chairman Steve Jobs unveiled the iPod on October 23, 2001. The iPod was the size of a deck of cards with a small screen, scroll wheel, and center select button. Its hard drive could hold five gigabytes, or approximately 1,000 songs. The computer magazine *PCMag* tested the iPod and reported that they had transferred 523 songs to the device in five minutes.[2] The original iPod was compatible only with Apple's Macintosh (Mac) computers. But the second-generation model, released in 2002, worked with PCs that were running the more popular Windows operating system. The number of people listening to portable digital audio devices grew.

The first iPod weighed only 6.5 ounces (184 g) and had a 10-hour battery life. Apple marketed the device with the slogan "1,000 songs in your pocket."

REBEL MEDIA

Because podcasts were downloaded and not broadcast over the airwaves, they were not subject to the same regulations as radio programming. The US government's Federal Communications Commission set these rules. If a song contained an obscene word, radio stations were required to play an edited version of the song. This rule didn't apply to downloaded podcasts. Yet it wasn't the fact that they could swear that drew early creators to podcasting. "The important thing is that you can be natural and not worry about what you're going to say," said Jim Kloss of the *Whole Wheat Radio* podcast.[3]

THE BIRTH OF PODCASTING

It wasn't long before people started recording their own audio to share online. In 2004, Ben Hammersley wrote an article for the *Guardian,* a British newspaper. In "Audible Revolution," Hammersley discussed how the internet and inexpensive recording software made it possible for people to create their own radio-style programming. Hammersley suggested several names for this kind of programming. One of the names was podcasting, a term that combined the words "iPod" and "broadcasting."

One person who became involved in this audio revolution was Adam Curry, a host for the cable channel MTV. Curry was already blogging and planned to add videos to his blogs. But software engineer Dave Winer convinced him that people would be more excited if they could download radio programs to listen to at their leisure. People could create audio programs and share them online, but they needed a way to distribute those programs.

For Curry, an existing technology called RSS (really simple syndication) was the answer. RSS uses a file format that computer programs can read. Programs can check the RSS file, also called a feed, to find newly updated programs. By 2005, Curry and other developers had created iPodder, a program that uses RSS files to download radio shows onto an iPod or other MP3 player.

To show how useful iPodder could be, Curry developed a podcast called *The Daily Source Code*. It ran until 2013 and had up to 500,000 subscribers.[4] Curry recorded some of the show's episodes in his car. He wasn't the only one to see the many possibilities of podcasting. Media companies such as the British Broadcasting Corporation (BBC) and the Canadian Broadcasting Corporation (CBC) soon branched out into podcasting. Podcasts could be just about anything, from product reviews to random recorded sounds.

APPLE CONTINUES TO LEAD

In 2005, Apple added a podcast directory to iTunes. The directory included more than 3,000 free audio podcasts and was one of the largest podcast directories at the time.[5] Apple users could download podcasts from established publishers such as ABC News, BBC, Disney, and National Public Radio (NPR). But in order for podcasting to really take off, people needed an easy way to make their own podcasts.

> **"Podcasting is the next generation of radio, and users can now subscribe to over 3,000 free Podcasts and have each new episode automatically delivered over the Internet to their computer and iPod.[6]"**
>
> ***—Steve Jobs, Apple, 2005***

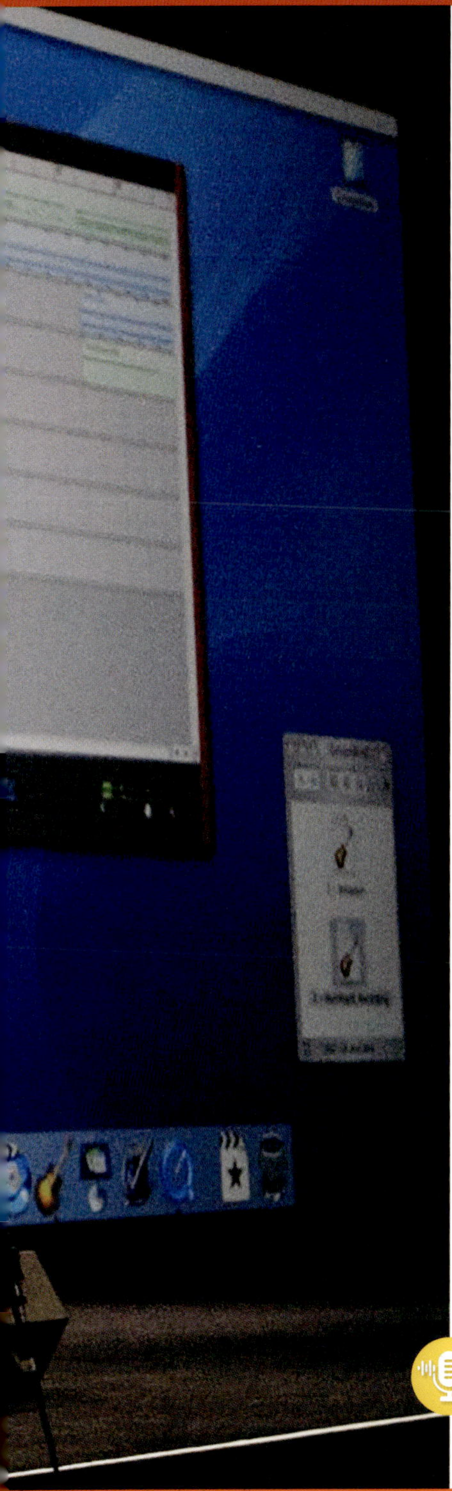

That came on January 10, 2006, when Apple announced updates to the audio program GarageBand at the Macworld trade show. This event focused on the latest developments for the Mac computer. GarageBand is a Mac program that enables users to create their own audio recordings quickly and easily. During a keynote speech, Steve Jobs explained that the new podcast studio software included 200 sound effects, 100 jingles, and automatic ducking.[7]

Jobs demonstrated the software in front of the audience, creating a humorous recording. "Hi, I'm Steve and welcome to my weekly podcast, Super-Secret Apple Rumors," Jobs said. "I've got some pretty good sources inside Apple, and the next iPod is going to be HUGE, an 8-pounder with a 10-inch screen . . . See you next week."[8] With this demonstration, Jobs showed everyone present that by using GarageBand, they could have a podcast episode ready to post in minutes.

As the number of podcasts grew, so did interest in this new form of media. In 2008, Edison Research reported that 37 percent of Americans surveyed had heard of podcasts.

The first version of GarageBand was introduced at Macworld 2004. Steve Jobs, *right*, invited musician John Mayer, *second from left*, to help demonstrate the new software for the audience.

By 2009, Edison Research reported that 43 percent of Americans surveyed had heard of podcasts.[9]

Podcasts grew steadily in popularity, but they didn't become mainstream until 2014, when the podcast *Serial* was released. In this true crime podcast, journalist Sarah Koenig followed the trial of Adnan Syed, who had been accused of killing his girlfriend in 1999. CNBC News reported that the podcast averaged 1.5 million downloads per episode.[10] In 2014, it became the first podcast to win a Peabody Award for excellence in storytelling. These awards are given to creators working in TV, radio, and digital media.

As more people started making and listening to podcasts, companies worked to gain bigger shares of the market. Spotify increased its reach in 2019 when it purchased the podcast studio Gimlet Media. That same year, Spotify purchased Anchor, a platform for creating podcasts, and the podcast studio Parcast. These purchases made Spotify a leading podcast platform.

In 2021, Apple changed things again with the addition of Apple Podcasts Subscriptions. The new service allowed podcasters to charge for premium content using Apple's payment system.

ONE BILLION SERVED

In July 2013, Apple announced that it had surpassed one billion podcast subscriptions. These subscriptions were divided among 250,000 different podcasts and eight million episodes. The shows were produced in a total of 100 different languages. Some podcasts performed better than others. NPR was one of the most popular podcast producers. By 2013, NPR had more than 25 million podcast downloads monthly.[11]

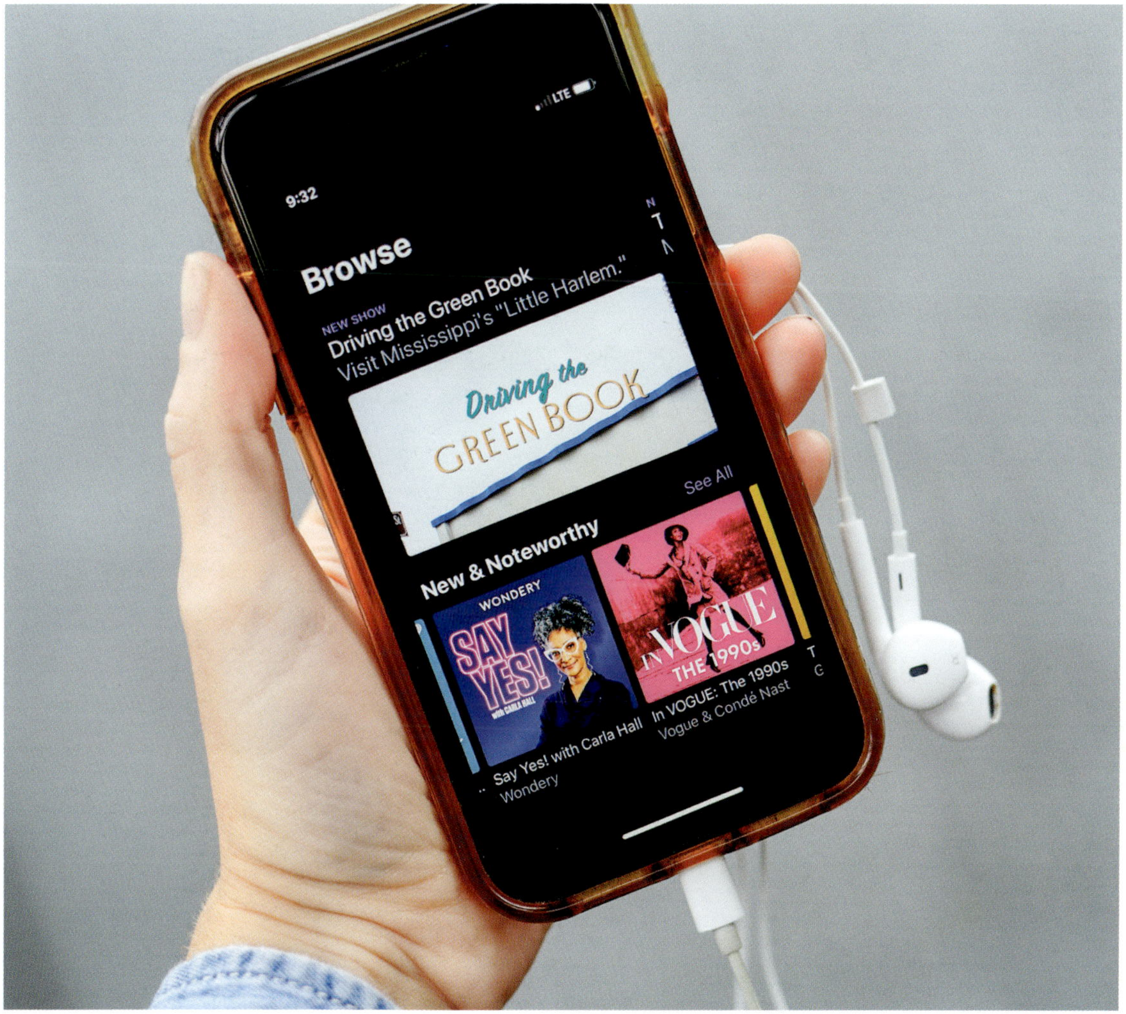

Apple Podcasts comes preinstalled on Apple products. Users can browse through millions of free podcasts or subscribe to access exclusive content.

This content might include ad-free episodes, early access, or exclusive access to a new podcast series. Among the brands that took advantage of this program were NPR, the *Los Angeles Times*, and Sony Music Entertainment.

However, there were some problems with the new service. Podcasters had to use an online dashboard called Apple Podcasts Connect, and the interface for the site was buggy. Users also encountered problems. Sometimes new podcast episodes didn't appear in their libraries. For other users, previously heard episodes poured into their libraries, and suddenly they could no longer sync across devices.

Podcasters complained that the system also required more work on their end. To charge users for only specific content, podcast developers had to upload content in multiple places. And Apple was slow to process uploads. For news content, a 24-hour delay could mean that content would be dated by the time it became available to listeners. Delays and an increased workload were especially difficult for podcasters who worked alone or on small teams.

PODCASTS TODAY

Podcasts are usually on-demand audio recordings. But some podcasters also choose to do video podcasts, which involves recording a video of themselves making the podcast. Then they can release both the regular audio version of the podcast and the

CHRISTOPHER LYDON

Christopher Lydon is an American journalist, TV personality, and broadcaster. In the 1990s, he hosted the radio show *The Connection*. Guinness World Records credits Lydon as the first podcaster, alongside software developer Dave Winer.

Winer was one of the first people to develop online blogging technology. He then developed a method of audio blogging. But he needed someone to be the voice of the audio blog. Lydon, with his broadcaster's voice and interviewing skills, was a prime candidate. In July 2003, Lydon and Winer met at Harvard University to record the first audio blog. It was an interview made using a voice recorder and laptop.

Lydon used Winer's technology to post the interview to his blog. It was the first episode in what became the longest-running podcast, *Open Source*. In 2023, Lydon was still hosting the podcast. He believes podcasting is open to everyone, which makes it a valuable form of media. "Anybody can be a podcaster," he said. "You can record your songs, your deepest thoughts, your silliest thoughts, your sonnets that you've been writing for years—you can spread it to the world."[12]

In *Open Source*, Lydon explores topics such as current events, culture, politics, and the arts.

PERSONAL AUDIO

In 1996, the company Personal Audio filed a patent to protect the technology it used to send out audio recordings of magazines on cassettes. In 2012, the company updated its patent to include internet distribution. Personal Audio later sued various podcasters for distributing audio content online. It said these podcasters had violated its patent and needed to turn over part of their podcasting income. The Electronic Frontier Foundation contacted the patent office to show that the CBC had serialized content before Personal Audio received its patent. Personal Audio was not the first to serialize content online, so no one owed it money.

video version. Release schedules vary, with some podcasts distributing a new episode each day and others putting out episodes weekly or monthly. Podcasts cover a wide variety of topics. Some, such as *Missing in the Carolinas*, explore true crime. Others, such as *Southern Gothic*, discuss hauntings, legends, and folklore. NPR produces the science podcast *Invisibilia*. Among the podcasts for foodies is *Gastropod*, which combines food science and culinary history.

In a podcast, each individual recording is an episode. Some podcasts follow a consistent format. Every episode might be a conversation between two hosts or an interview with an expert. Other podcasts feature a variety of formats. They might include both interviews and news summaries while remaining focused on one topic.

Podcast episodes typically open with an intro, in which the host welcomes listeners to the podcast and briefly reviews what the podcast is about. For example, the host might say, "This is the podcast where we discuss all things Broadway." Following the intro is the body of the episode. The podcast usually ends with the outro or closing credits. In the outro, the host thanks everyone who helped make the episode and thanks listeners for tuning in. The host may also tell people where to find more information about the podcast.

Podcasters upload each episode's audio file to a server connected to the internet. Listeners use their favorite podcast app to subscribe to the podcast's feed. The app automatically downloads new episodes as they become available. Some apps, such as Spotify and Apple Podcasts, include directories where people can search for podcasts that interest them. Other popular podcast apps include iHeartRadio, Google Podcasts, Amazon Music, Castbox, Overcast, and Pocket Casts. People can also listen to podcasts through their web browsers.

Podcasts have become a popular form of media. Millions of podcasts exist on all sorts of topics. But compared with forms of media such as television and radio, it is still relatively easy for newcomers to make and distribute their own podcasts. Today, people can use their phones to listen to podcasts wherever they go. Many podcast listeners are interested in a wide variety of topics and are looking for entertaining takes on what intrigues them.

BLACK DIAMONDS
SiriusXM
NEGRO LEAGUES BASEBALL MUSEUM
BLACK DIAMONDS
sxm

DEVELOPING A PROFESSIONAL PODCAST

Many professional organizations and businesses produce podcasts. Podcasts are a great way for organizations to make their messages more accessible to the public. Sometimes organizations use podcasts to strengthen ties with their existing audience, which works especially well for radio stations. "Obviously, podcasting is a very natural fit for our audio focus," explained Zach Brand, NPR vice president of digital media. "Our audience loves listening to our content through podcasting, and we love delivering it. We like to engage our audience on whatever platform they choose to listen."[1]

Other organizations find that podcasts expand their reach, which is important for nonprofit and charitable organizations. The Georgetown Institute for Women, Peace and Security (GIWPS) researches the roles women play in peace work and economic development. Before the GIWPS started podcasting, it shared information about its mission, studies, and findings through dense academic documents. Only a small number of scholars read these written materials. The GIWPS wanted

Some professional organizations use podcasts to educate listeners. In 2021, the Negro Leagues Baseball Museum teamed up with broadcasting company SiriusXM to create the *Black Diamonds* podcast. It explores the history of Black baseball stars.

FILLING THE WELL

Many types of nonprofits have podcasts. Arts Midwest supports arts organizations and communities throughout the midwestern United States. It hosts the podcast *Filling the Well*. This podcast seeks to energize and inspire artists and leaders. One episode focuses on community-centered agriculture and art. Another discusses the Indian Arts and Crafts Act and how critics believe it is failing American Indian artists by not promoting their careers and work. One episode explores how nature helps Sam Miltich, a jazz guitarist with schizophrenia, reduce his symptoms.

to make more people aware of the life-or-death circumstances many women around the world face. It saw the courage that these women demonstrated and wanted to share this bravery with a larger audience.

The GIWPS launched its podcast, *Seeking Peace*, in December 2020. It explores the work of activists and women who help bring peace to their communities. The podcast did more than just educate the global community. The women profiled on the podcast received more local support. One episode profiled Yemeni activist Wameedh Shakir. She was surprised by the posts, shares, and comments that her interview generated from Yemeni women, men, youth, and journalists.

Organizations with a tight local focus can spread their messages to a national audience through podcasting. One example is the Lower East Side Girls Club in New York City. This organization's mission is to break the poverty cycle by giving girls skills that they can use to get jobs. In 2015, the Girls Club started the podcast and online radio station *Where Girl Radio Lives NYC* (*WGRL NYC)* as a way for girls to learn about radio journalism and audio engineering. This youth-led programming has brought the group's work before a national audience, raising both funding and awareness.

BRAINSTORMING

It may seem like a simple task for an organization to develop a podcast. Someone decides to podcast about the group's goals, talks it over with a few people, and then gets to work. But the process is much more complicated than that, especially with a large organization. One person may come up with the initial idea for an organization's podcast. But it takes support from people across the organization to get the podcast off the ground.

Tony Cohn was a brand marketing specialist at the Smithsonian Institution, a large group of government-created museums, when he first came up with the idea that became *Sidedoor,* a Smithsonian podcast. As a marketing professional, it was Cohn's job to help the Smithsonian show a consistent image to the public. The Smithsonian describes itself as the world's largest museum, research, and educational complex. Its goal is to preserve US heritage, discover new things, and share its collections and findings with the world.

In 2016, Cohn was riding the subway to work in Washington, DC. He was listening to a podcast and looking out the window. He wondered why the Smithsonian didn't have a podcast. Scrolling through Apple's podcast directory, Cohn found that several

Geology, Gems & Minerals
Geology, Gems & Minerals
Fossil Hall
Fossil Hall

individual Smithsonian museums had podcasts, but the larger organization did not. To show the public that the museums were all working toward the same mission, Cohn thought the Smithsonian needed a podcast that represented the whole organization.

When Cohn got to the office, he told a colleague about his idea for a Smithsonian podcast. She liked it, and soon a group of people were discussing it. Cohn and a coworker borrowed microphones and created a sample to show other Smithsonian staffers what they wanted to do. They soon gained support from the Smithsonian's digital team. The team assigned Megan Detrie as the series producer. She and Cohn produced the first episode of the *Sidedoor* podcast, which was titled "Tech Yourself." The episode brought together three stories about cell phones, computer simulations, and time.

Over several months, Detrie and Cohn created the first season of the podcast. In the end, more than 100 Smithsonian employees, including animal keepers, artists, curators, and security guards, had participated in the podcast.[2] The first

The Smithsonian Institution includes 21 museums, such as the National Museum of Natural History. The Smithsonian Podcast Network offers a variety of podcasts exploring museum-related topics.

ONE OR MANY

Some organizations have only one podcast, such as the single-season *Welcome to My Garden*. It was produced by the Missouri Botanical Garden to celebrate its 150th anniversary in 2009. Other organizations have many podcasts. The Smithsonian Institution has nineteen.[3] The Library of Congress has produced fourteen.[4] The key to making multiple podcasts work is maintaining a tight focus or limited run so that each podcast features a unique topic. The Smithsonian's *Articulated* podcast, for example, focuses on the institution's oral history program. Its podcast *Collected* focuses on African American history topics.

season included stories about anthropology, panda conservation, the National Museum of African Art, and comedian Phyllis Diller. All of these topics involved the Smithsonian and its collections, museums, and conservation centers. In September 2023, *Sidedoor* was on season ten.

Seasons are one way to produce podcasts. A season might last months or years. During the season, episodes are released on a regular schedule. At the end of the season, the podcaster usually takes a short break. For listeners, seasons break content into manageable chunks. This way, someone who finds a podcast can browse episodes by season instead of trying to review hundreds of episodes.

BUSINESSES PODCASTING

It isn't just organizations and nonprofits that podcast. Businesses may decide to podcast as well. Businesses generally podcast because they want to get more customers, but they can't simply use podcasts to tell people to buy their goods and services. First, they need to draw people in. Mike Andes runs Augusta

LIZZIE PEABODY

Lizzie Peabody is the senior producer of the Smithsonian podcast *Sidedoor*. Her first podcast was *Your Story Here*, which she created and produced herself in 2016. The inspiration for *Your Story Here* came from a New Year's resolution. Peabody wanted to improve her interviewing skills, so she challenged herself to do an interview every day for a year.

She recorded the interviews on the microphone built into her iPhone 5 and edited them using free software. Her brother played the theme music for the podcast on his keyboard. Peabody shares these details on her website, letting people know that it doesn't take a lot of technology to create a podcast.

Peabody continues to use her interviewing skills in the production of other podcasts. A storyteller herself, she was a producer for *Story District Presents*. The nonprofit Story District encourages everyday people to tell their truths on stage at live shows known as Story District Presents. The podcast combines stories from these live shows and interviews with the storytellers themselves.

No matter which podcast Peabody is working on, her role as producer is to keep the podcast on track. She edits the audio and makes sure episodes are posted so eager fans can hear what happens next.

In addition to *Sidedoor*, Peabody has worked as a producer for the Smithsonian podcast *AirSpace*.

Many episodes of *Inside Trader Joe's* inform listeners about new or seasonal products. In 2023, episode topics included fall shopping lists, food hacks, and product testing.

Lawn Care Services. He has a podcast about building a small business. The exposure he received because of the podcast helped him expand his company.

In 2018, the Trader Joe's grocery store company created a five-episode podcast called *Inside Trader Joe's*.[5] It was so popular with customers that it became an ongoing series. Individual episodes cover a variety of topics, including party planning, shopping lists, and employee recommendations. Other episodes focus on things that customers might not normally hear about, such as how stores are stocked or why Trader Joe's sells private-label products. The popular podcast keeps listeners tuning in and helps them build a connection to the brand, making customers keep coming to the store.

Other companies create podcasts that have a looser connection to their business. One example is General Electric (GE), a multinational company that works in power, renewable energy, aviation, and internet services. GE created the podcast *The Message,* a serialized science fiction story in which scientists decode an extraterrestrial message. The scientists in the story use technology developed and sold by GE. Even though their industries are sometimes seen as serious and stuffy, GE's podcast is entertaining and engaging for listeners. The podcast challenges what people think about the company itself.

> **"Your customers won't just be purchasing your product or service, they'll be getting to know you, and becoming your fans. Your show provides a simple and effective way to celebrate and inform your audience.[6]"**
>
> ***—Steve Davis, creator of online podcasting tutorials***

Businesses can also use podcasts as a form of advertising. They can do this in several ways. One way is for businesses to place brief advertisements within the podcast episode itself. This could be as simple as mentioning the business in the podcast's intro or outro. "Happy Pet is brought to you by Hazelwood Valley Veterinarians," the

TEACHING TOOLS

***The Manuscript Academy* podcast focuses on educational topics for writers, from writing query letters to staying positive. It is produced by literary agent Jessica Sinsheimer and media professional Julie Kingsley as part of Manuscript Academy. This organization teaches writers about the publishing industry, offering courses and consultations with agents and editors. The podcast is free, but membership in the academy requires a fee. By paying the fee, writers gain access to classes, critiques, and more. The podcast gives Sinsheimer an opportunity to share information about upcoming Manuscript Academy classes and persuade listeners to enroll.**

host might say. "Bring your pet in and get a 10 percent discount when you mention listening to this podcast." Or the advertising could consist of product placement, such as when scientists in *The Message* use GE products to solve mysteries.

Whether their purpose is to entertain or teach, podcasts are usually most successful if they give the listener something of value. This is especially true of podcasts produced by businesses. If a lawn care company discusses only problems that require a lawn care company to solve, people might see it as self-serving. But a lawn care company that tells listeners five things they can do to maintain a healthy lawn, as well as when to bring in a professional, is more likely to be seen as helpful.

Podcasting can also help a business or organization build a community. This happens when listeners comment on the podcast and start discussions. It can lead to back-and-forth conversations between the listeners and the company or organization, along with conversations among listeners themselves. An online community can also help build a network. Engaged listeners may talk about the business, organization, or topic with people who

One of the *San Francisco Chronicle*'s podcasts is *Total SF*, hosted by columnist Heather Knight, *bottom right*, and culture critic Peter Hartlaub, *left*. It focuses on the people and places of San Francisco.

don't listen to the podcast, extending the organization's reach. This can be vital in getting the word out about a need or opportunity.

A network is also a great way for a business or organization to gather information and learn about its customers. If a business isn't sure about what products customers might want to see next, it can ask its community. The responses the business receives may not guarantee sales, but they can keep the business from spending time and money creating products their customers aren't interested in. Podcasts can help businesses and organizations grow while giving something valuable to their listeners.

HISTORY

CHAPTER **FOUR**

RECORDING A PROFESSIONAL PODCAST

Once an organization chooses an idea for a podcast, it gathers a team of people to develop the project and record episodes. Organizations that can hire large staffs have an edge in podcast creation. Instead of having just one person learn how to do various jobs, organizations and businesses may create a podcast team. The first person on board is often the producer.

The producer has a wide variety of responsibilities, including coordinating the jobs of other people working on the podcast. The producer makes sure all episode ideas are in line with the podcast's theme. They also keep track of the budget. This means the producer notes how much money is spent and makes sure it is no more than the amount set aside for the show. Expenses might include paying podcast writers or covering travel costs for guests.

If the podcast is produced by a business, the producer communicates with both the business owner and the podcast team. They maintain the schedule. They also make sure tasks are completed so that each episode will be released on time.

Some podcasts, such as History Colorado's *Lost Highways*, have multiple producers. Senior producer Noel Black leads a team of producers who work on the podcast.

I am specializing down in particular subject areas, which I would also recommend doing, because then people come to you for your specific expertise, and you get more of the work that you really enjoy.[1]

—Josie Colter, podcast producer

The producer is often the person who checks the podcast's metrics. This means the producer checks to see how many people have downloaded each episode. Depending on where the podcast is hosted, the producer may have access to even more information.

Apple Podcasts gathers data about how long people listen to each episode. This can be useful information for podcasters. If a podcast is one hour long and people listen to it for only fifteen minutes, the podcaster knows it might be a good idea to produce shorter episodes. Apple Podcasts also notes if listeners skip the intro to go to the show. Metrics include ratings that show how much people like the podcast too.

Some producers specialize in podcasts centered on a single theme. But others work more broadly. Annie Galvin produces the current events podcast *The Ezra Klein Show* for the *New York Times*. She has done episodes on music, politics, and LGBTQ+ history. Galvin also produced the *Public Books 101* podcast for the magazine *Public Books*. Each season presented scholars discussing a particular topic, such as the internet or the value of novels.

JOSIE COLTER

Josie Colter is a podcast producer and marketing consultant who has worked on TedxLondon's *Climate Curious* and Gen.T's *Up to Speed with UBS*. Her initial experience was in marketing a podcast. When the podcast's producer went on maternity leave, Colter asked to do the job. As other clients came to the company for help developing podcasts, the company recommended Colter. Her producing skills continued to grow.

Producing podcasts wasn't a job Colter considered as a student. She listened to music on her iPod but didn't even know that producing podcasts was a job. She liked to try out new platforms and learned about media in a hands-on way. By working in advertising, Colter learned to build a brand, which has helped her increase the reach of the podcasts she works on.

Colter is curious to see where podcasting will go in the future, especially with the creation of niche shows appealing to specific audiences and innovations in podcast formats. She suspects there will be more video podcasts. But she still believes that audio is a good way for people to build their brands and reach out to others.

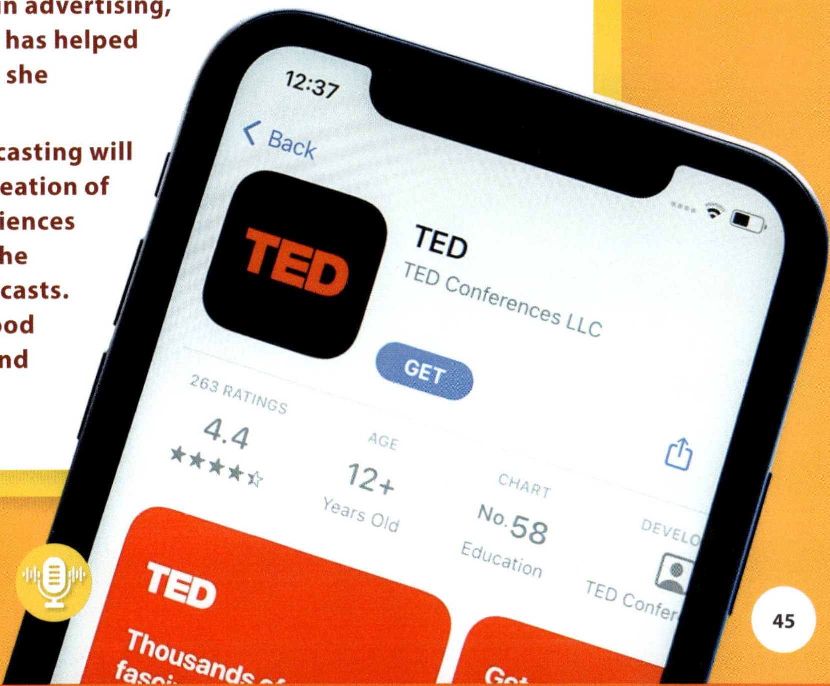

In addition to her work as a podcast producer, Colter serves as head of engagement for TedxLondon.

RESEARCH

There is a lot of research involved in making a good podcast, especially if the podcast is based on interviews. The host needs to know about the person they are interviewing. They must know about the topic of the podcast too. Because they need to know enough to ask good questions, they must review other interviews with the person to avoid asking the same questions other hosts have asked. Research may also be required to identify topics for future episodes and seek out experts to interview. This research can be done by the producer, writer, host, or a paid researcher.

THE WRITER

Another person who helps create a podcast is the writer. The writer outlines or writes the episodes. Some podcasts are recorded from a full script, meaning that all dialogue is written out beforehand. Bigger podcasts that release episodes frequently may employ several writers.

In addition to writing the actual episodes, the writer creates additional content for the podcast. This may include show notes that describe the episode, along with a transcription of what was said on the podcast. Transcriptions are often found on the podcast's website and are especially important if the host improvises instead of reading a detailed script. The writer also creates any articles or blog posts related to the podcast. These materials are usually posted on the podcast's website.

The writer may also need to prepare advertising materials, including a press kit and an advertising or sponsorship kit. The press kit contains everything a journalist might need to know in order to write an article about the podcast. These materials include a list of episodes and release dates, along with information about popular people who have appeared on the podcast. An advertising or sponsorship sales kit provides detailed metrics about the podcast. A salesperson can use this to show a potential advertiser how many people listen to the podcast and what kind of reach an ad on the podcast would have.

THE HOST

The host is the on-air talent and voice of the podcast. This is the person who people listen to, so choosing a host with an appealing voice is a must. What that means depends on the podcast and its audience. For one podcast, the host might need to sound soothing. But for another, the host might need to sound firm and authoritative.

In addition to having the right voice, a podcast host must also have the right personality. Most hosts are outgoing, possess a good vocabulary, and have excellent public speaking skills. If interviewing people on-air is part of the podcast's format, the host must know how to interview someone. They must pay close attention to what the interviewee is saying and know how to show they are listening without interrupting. Hosts must also be

ADS WORK

Podcast ads work because the advertising goes out to a tightly focused audience. A person listening to a podcast about launching a small business is likely interested in the topic, along with the products and services that may help someone successfully start their own small business. One study from the media analytics company Nielsen showed that podcast listeners recalled a product from a podcast ad more than four times better than people who had seen a print ad.[2]

NPR'S POP CULTURE HAPPY HOUR
Casper
PODCAST LOUNGE
AT HIGHLINE STAGES
PUBLIC
ADIO
ERD

able to improvise questions in response to what the interviewee says. They should know how to take quick notes to hold a thought until the interviewee has finished speaking.

Some podcasts have only one host. Ezra Klein, for example, hosts *The Ezra Klein Show*. But other podcasts have multiple hosts. Josh Clark and Chuck Bryant are the cohosts for *Stuff You Should Know*. This podcast offers information on a wide variety of topics, from how toy testing works to how magnets work.

> **“Keep your questions clear and direct. . . [Make] sure that your questions are really clear and concise, and maybe ones that they haven’t heard a thousand times before.[3]”**
>
> ***—Sarah Li-Cain, podcast host, 2021***

THE AUDIO ENGINEER

The audio engineer works with the host to make the actual recording of the podcast. Part of this job involves checking that the recording levels are correct. This might mean making sure the audio is loud enough to be easily heard without sounding distorted. It also means that during an

NPR’s *Pop Culture Happy Hour* podcast is hosted by arts journalists, including Linda Holmes, *left*, Stephen Thompson, *second from left*, and Glen Weldon, *third from left*. Many listeners find their discussions insightful and funny.

JOBS CAN OVERLAP

Some large organizations have an entire department to handle podcast recording and sound editing. But other times, people do more than one of these jobs. Which team members take on multiple jobs depends on their individual skills and knowledge. For example, the host may do the intro, outro, and any ads that are not prerecorded. Other times, a voice-over specialist may tackle these jobs. Some podcast teams start out small, with just a few people handling multiple tasks. Over time, the podcast may grow and new staff members may be added to the team.

interview, both people are recorded at the same sound level so that one person's voice is not significantly louder than the other. These things can be adjusted after the recording is made, but getting a good recording means it will take less time in postproduction to edit the podcast. The audio engineer also makes sure no background noises such as traffic or a barking dog can be heard on the recording.

Professional podcasts are often recorded in a sound studio, which makes the audio engineer's job much easier. Businesses that provide professional studios often have several studios set up in a single location. Each studio contains equipment for recording and producing the podcast, including microphones, monitors, speakers, cables, computers, and soundboards. Recording studios also have soundproofing. This prevents outside noises or echoes from disrupting the recording.

Sometimes the audio engineer takes care of postproduction tasks as well. Other times these tasks are the job of a postproduction audio editor. During postproduction, this person adds sound effects and music to the podcast, along with the intro and outro. They also make sure any prerecorded ads are placed in the recording.

Audio engineers may use soundboards, mixers, and computer software to adjust and edit podcast recordings. They may also assist during recording sessions, setting up equipment and telling the hosts how far away to be from the microphone.

Along with podcast artwork and logos, graphic designers may design ads and posters for a podcast. Food magazine *Bon Appétit* produces the *Dinner SOS* podcast. Its podcast artwork features a cooking pan on fire.

Audio engineers often work in both the music and podcasting industries. Josh Giunta, an assistant professor at Berklee NYC, is the head engineer at Manhattan's GSI Studios. In addition to recording musicians such as Yebba and Questlove, he has recorded podcast episodes with John Mayer and Malcolm Gladwell.

THE GRAPHIC DESIGNER

Some podcast teams hire graphic designers. Sometimes graphic designers are employees of the production company, but in other cases they may work on a freelance basis. Before the podcast launches, the graphic designer creates the podcast art. This artwork will be used on the show notes, website, and various directories that potential listeners use to find podcasts.

The graphic designer may also make individual graphics for each episode. These graphics generally include episode-specific artwork, along with the name and number of the episode. If the podcast is based on interviews, the graphic designer may use headshots of the guests. The graphic designer also creates artwork to be used on social media for publicity purposes. These pieces may be built around quotes, statistics, or other material related to the episode. When working on a podcast for a business, the graphic designer may also create images for the company's website, blog posts, or publications.

DEVELOPING AN INDEPENDENT PODCAST

Organizations and businesses are not the only ones who create podcasts. Individuals make podcasts about a wide variety of topics too. Carissa and Jason Weiser podcast about folklore on *Myths and Legends*. Kendra Adachi created *The Lazy Genius Podcast,* in which she gives advice about home and work life and assures listeners that they don't have to do all things well.

When an individual creates their own podcast, they have a lot of decisions to make. It all starts with choosing a topic. Some people might be tempted to choose a topic because they think it will be popular. And it's easy to tell which podcasts are popular. The Google Podcasts "Popular & Trending" list includes *Stuff You Should Know*, the *New York Times* podcast *The Daily*, and the news podcast *Dateline NBC*.[1]

The problem with picking a topic just because it is popular is that it might not be something the podcaster is passionate about. If a person who doesn't care about politics tries to make a podcast about politics, it is going to show. If they find science tedious, their listeners won't stick around because they will be able to tell

Former radio broadcaster Randy Lane, *left*, noticed there weren't any podcasts about the history of Waco, Texas. He teamed up with history professor Stephen Sloan, *right*, to create the *Waco History Podcast*.

SATURATION

If a person creating a podcast wants to monetize their work, it is important to make sure that they will have enough listeners. If the market is saturated, this may be difficult. Topics such as health, fitness, and lifestyle are popular, but there are already a lot of podcasts on these topics. A new podcaster will need a unique angle or expertise to break into the topic. A person with an engaging voice who talks about affordable or thrifted fashion options might find listeners. But someone who wants to talk about fashion in general might have a harder time.

that the podcaster doesn't care. A podcaster who doesn't care about their topic will likely sound fake or insincere to listeners.

The other problem with not caring about a topic is that the podcaster will probably not enjoy the experience. Podcasting takes a lot of work. Podcasters may be in school or have jobs. They may have families, hobbies, and daily responsibilities. In order to fit in time to create a podcast, a podcaster must really be interested in their topic.

Businesses that help people create podcasts suggest that would-be podcasters look at their Google search histories. Maybe they've done online research about the possibility of a Mars colony. Or perhaps they have read about the migration of monarch butterflies. People can also look to their hobbies or interests for inspiration. Maybe they are interested in manga or drones. Anything that fascinates someone is a potential topic for a podcast.

Choosing a topic is only the beginning of the process. Once a person has selected a topic, they must decide how broad or narrow to make their podcast. For example, drones are a broad topic, while drone photography is a narrow topic. To find out if a topic is too

RENEE ROBERSON

Renee Roberson is an author who moved to North Carolina when she was 13 years old. Even then, she was an avid reader with a love for mysteries. She later became a journalist and started listening to podcasts in 2018.

In 2019, Roberson attended MurderCon, a conference for authors interested in writing about crime. After the conference, she started her own podcast, *Missing in the Carolinas*. Roberson does the research and writing for the podcast. She also produces the show independently.

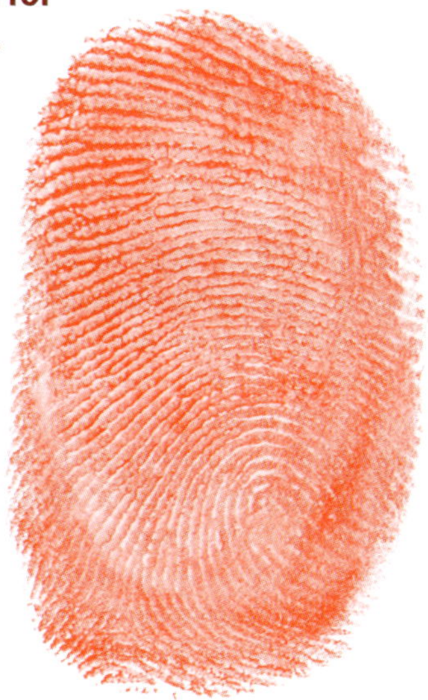

Roberson often focuses on forgotten cases. Some episodes are about specific crimes, such as the Hudler Carolina Tree Farm murders. Others focus on a broader category, such as murders involving sunken cars. Many are cold cases with no leads, but other cases featured on the podcast have gone cold because reporters ignored them and never featured them in the news.

In addition to talking about crimes, Roberson aims to teach listeners situational self-awareness. She wants people to understand how important it is to know what is going on around them, which can help them to avoid becoming victims.

In her podcast, Roberson explores both cold cases and cases that have been solved.

broad, the person can search for it among current podcasts. What do other podcasters discuss when they talk about drones? This is a balancing act because a topic that is too narrow may be quickly exhausted, while a topic that is too broad may make it difficult to attract an audience.

The final test for a podcast topic is to brainstorm episode ideas. The company Buzzsprout helps people create podcasts. According to advisers at Buzzsprout, five ideas won't be enough for a season. Ten ideas is a good start.[2] If the podcast is interview-based, the podcaster should brainstorm potential guests. The guests don't have to be famous, but they do need to be interesting. Not every idea will become a podcast episode, but brainstorming ideas is a good way to tell if a topic is too narrow to work in the long term.

> **"Maybe, if you're doing interview style, it's 30 minutes to maybe an hour or even a little bit longer, so that you can really draw out the right content from your guest.[3]"**
>
> ***—Heather Torres, Think Media, on podcast lengths***

CHOOSING A FORMAT

The next major decision involves choosing a format for the podcast. If the subject is something that could inspire good discussions, the podcast might involve two or three cohosts who will converse about the topic. This is the format for *Stuff You Should Know*.

A podcast can also be interview-based. In this format, the host asks experts a series of questions. That's how host

Stage actress Millie Brooks runs the podcast *Me, Myself & Millie*. Recording from her home, she discusses her experiences with infertility and interviews others about pregnancy and parenthood.

Many podcasters choose titles that are short, sweet, and easy to remember. For podcasts with lighthearted or fun topics, podcasters may choose a clever pun or play on words for a title.

Mark Asquith teaches everything from branding to media launches in the podcast *Straight Talking Guide to Launching Your First Product*.

Another possible format is a solo podcast, in which a host discusses a topic alone. This format works well for current events and science topics. It is also a good format for reviews of books or restaurants. Movie review podcasts, such as *Now Playing* and *In Review,* are popular with listeners.

Storytelling or radio theater is another option. This involves the host reading or telling a fictional story during the podcast. Sometimes a cast of actors might read different parts of the story. These dramatic podcasts, such as *The Message,* often include sound effects or music to heighten the tension or add a bit of comedy. Nonfiction can also be delivered in a storytelling format. In a travel podcast, the host may relate what they witnessed or experienced while traveling. Other podcasters might describe a historic event as if the audience was witnessing it. True crime podcasts often use this format. One popular example of a nonfiction podcast with a storytelling format is *S-Town*, in which producer Brian Reed talks about his experience investigating an alleged murder.

There are several things to consider when selecting a podcast format. One is the goal of the podcast. A podcast might be intended to entertain, educate, or challenge people. The goal of a podcast must work with its format. For example, storytelling formats can be both entertaining and informative. But a serious topic delivered as a news story will be more educational than entertaining.

Another factor to consider is the length of the podcast. Some podcasts are more than an hour long, while others are as short as 15 minutes. The length of an episode must suit the podcast's topic and format. Interviews are often in-depth explorations of a subject and may require longer recordings. Topics such as product reviews can often be handled in a shorter amount of time.

HOW MANY MINUTES?

Buzzsprout checked the lengths of the podcasts on its platform and found that the most popular show length was 20 to 40 minutes, with 31 percent of all podcasts fitting into this slot.[4] Because many people listen to podcasts in the car, on the bus, or while exercising, shorter programs are more popular. Shorter podcasts are also quicker to produce. It's easier to stay on topic because there isn't time to waste on tangents.

PODCAST ARTWORK

Podcast artwork is the square image that potential listeners see when they look up a podcast on Apple Podcasts or other directories. The artwork generally includes the name of the podcast, so a short name is ideal. The image must also be easy to recognize on both phones and laptops. Most podcasters wait to make podcast artwork until they are done fine-tuning the podcast's name.

People should also consider their podcast's differentiator, or what makes it unique. Karina Longworth makes her podcast unique by focusing on a niche. Her podcast *You Must Remember This* is about forgotten Hollywood history from the 1900s. Sometimes a podcast's topic makes it unique. Other times, the format is unique. Often, it is a combination of the two.

THE DETAILS

Once the podcaster has narrowed down the subject and format, it is time to come up with the podcast's title. This is the first thing that potential listeners encounter on podcast directories, alongside the podcast art, so it must stand out among other titles and be easy to recognize. Some podcast titles are abstract but interesting. *Twenty Thousand Hertz* is about easily recognizable sounds. The podcast *99% Invisible* focuses on the world of design and architecture. Someone who encounters either of these titles for the first time will not know exactly what the podcasts are about, but they may be intrigued enough to click on them and learn more.

Other podcast titles are highly descriptive. These include *Stuff You Should Know*, *Missing in the Carolinas*, and *Stories Podcast*. The first podcast is about random interesting topics. The second focuses on missing persons in the Carolinas. The third features bedtime stories.

A podcast title needs to do more than catch a listener's attention. It should also match the podcast's tone. A humorous name would not be a good match for a serious topic, such as politics, unless it matches the podcaster's joke-filled approach.

Once a podcast title has been chosen, it is time for a search. Podcasters must make sure that no one is already using the name they have selected. It may take several attempts to find a good name for a podcast that fits the show and is not already being used.

The last detail to work out before recording begins is the podcast's publishing schedule. The company Buzzsprout recommends that new podcasters should release one episode every week.[5] Frequent episodes attract a larger audience.

"Be very careful about the name you choose, and try as well as you can to find a unique one.[6]"

—Colin Gray of thepodcasthost.com

HEIR TO THE

CHAPTER **SIX**

RECORDING AN INDEPENDENT PODCAST

One of the most intimidating aspects of starting a podcast is figuring out what recording equipment to use. A new podcaster can go as basic as they want, using their phone to make the recording. Although it can be hard to get a good recording using a phone's built-in microphone, basic headphones are all that is needed. Headphones allow podcasters to hear how they sound to other people. While recording, podcasters can change how they speak. This might mean changing their volume or tone.

Headphones can also improve a podcaster's microphone technique so that the mic doesn't pick up certain noises, such as the popping from "p," "b," or "k" sounds. "You can use any set of headphones . . . as long as it's got the microphone connected to it because that's going to level up the audio just a bit from the one that's actually on your phone," said Heather Torres of Think Media.[1]

Many podcasters start out by using their laptops, a headset, and a microphone to record. But selecting a microphone isn't easy. The first step is understanding the

Some podcasters have recording stations and equipment set up in their homes. Jimmy "Mac" McInerney, for example, records episodes of the Star Wars podcast *Rebel Force Radio* in his basement.

difference between a dynamic microphone and a condenser microphone. Condenser microphones are very sensitive and can pick up the subtleties and nuances in a person's voice. Because of this, these microphones are often used to record music.

> **Find a small and soft place to record. I know that might sound kind of funny, but somewhere like inside your closet is perfect because it's a small space and you are surrounded by soft items like your clothes that will help reduce the echo.[2]**
>
> ***—Katie Steckly, podcaster and digital entrepreneur***

But condenser microphones pick up many other noises as well, which can be problematic if a room isn't soundproofed. This makes it difficult for podcasters who record their podcasts at home, where the microphone can pick up sounds made by other people. A dynamic microphone is better at picking up lower frequencies but is generally less sensitive. This type of microphone tends to work well for most independent podcasters.

Once a podcaster has chosen a microphone, they must learn about the two common types of connectors that microphones can have. These are XLR and USB connectors. A microphone USB connector is the same as the USB connector used to connect devices to computers or chargers. USB microphones are less expensive and record high-quality sound. This makes them a good choice for most new podcasters.

Podcasters can use microphones of all shapes, sizes, and prices. Some podcasters also purchase accessories for their microphones, such as mic stands, boom arms, or shock mounts.

An XLR microphone uses a three-pronged connector. These types of microphones and connectors are used in the professional audio industry for things such as recording music. Most professional recording gear is built to be compatible with XLR microphones, but these microphones are expensive. They must also be used with a sound mixer.

A sound mixer is a piece of equipment used in professional recording studios. XLR cables from different microphones connect to the mixer. This allows the sound engineer to adjust the levels coming from each microphone. This means a quieter sound can be made louder while a louder sound can be pushed into the background. Podcasters may need to combine audio from phone calls, music, sound effects, and their own voices at the same time. This can sometimes be done on a laptop. But if a podcast has two hosts who each have their own microphone, then a mixer may be necessary. Sound engineers recommend this setup.

> **"You want to mention the things that give credence to your advice. This is specifically important if your podcast is centered around you and your expertise.[3]"**
>
> ***—Travis Albritton, Buzzsprout, on establishing credibility***

GETTING READY TO RECORD

Deciding how to record a podcast and which gear to use is important. But new podcasters must also think about preparing the first episode. Many beginners are tempted to start recording without writing a script first. Attempting to record without a script can result in a rambling podcast full of off-topic conversation. This kind of free-wheeling discussion may be a good fit for some podcasts. But a script is helpful when podcast creators want to make a professional, focused episode.

Some people write out the entire script, word for word. But the script doesn't need to be detailed. A simple bullet point outline can be enough and might fit on a single page. The script should begin with an introduction that summarizes the episode. Next it should list the first topic of conversation or first point that the podcaster wants to make. Each topic or point should have supporting points or data. Then there should be a transition to the second topic. After that topic, there might be another transition into the final topic. Finally, the script will end with a conclusion or wrap-up.

SCRIPT TEMPLATES

One thing that can make writing a podcast script easier is a template. Whether a podcaster is scripting a conversation between two hosts or a tutorial that teaches listeners how to bake a cake from scratch, a template can give the podcaster guidance and confidence. It can also help podcasters remember what information they need to include. Podcasters can google "podcast script template" or check out templates on hosting services such as Buzzsprout, Riverside, and Ausha.

When a podcaster finishes scripting the first episode, they should also create the intro and outro. The intro should include several things. First, the podcaster must welcome listeners to the podcast, say the podcast's name, and introduce themselves. For example, the podcaster might say, "Thank you for listening to *All About Magic*. I am your host,

John Doe." Then the podcaster may share their credentials or what makes them qualified to speak on the podcast's topic. They might tell listeners, "I've been practicing magic for the last three years." In the outro, the podcaster should thank their listeners, remind them to subscribe, and thank anyone who helped make the podcast. This is also a good time to remind people to visit the podcast's website for show notes, special offers, and additional information.

Before recording the full episode, the podcaster should record a sample to make sure everything is set up correctly. A test recording helps verify that the recording location results in good audio quality. The ceilings in recording studios are covered with sound-absorbing materials. This is because sound bounces off hard, flat surfaces such as walls, ceilings, windows, and doors. Recording in a room with hard, flat surfaces can lead to a recording that has poor sound quality or strange echoes.

A podcaster who doesn't have their own recording studio can avoid this problem by recording in a room that has upholstered furniture, curtains, and rugs. Podcasters can also

Podcasters Tess Barker, *left*, and Barbara Gray, *right*, recorded episodes at the Earwolf podcast studio in Los Angeles, California. Podcasters may need to rent or pay to use space at a recording studio.

HD 280 PRO

buy foam acoustic panels that absorb sound and reduce echoes. They can use these acoustic panels to set up their own at-home recording studios.

PODCAST EDITING

Once a podcast has been recorded, it is time to edit the recording. This step helps improve the podcast's sound quality and conversational flow. Some people record the intro, podcast, and outro in one take. If they are happy with the quality, they can post their recording without any editing. But most podcasters do some editing before posting their podcasts. They might cut out long gaps of silence or add music and effects. Podcasters can use several different computer programs to record and edit audio files.

Audacity is a free program available for both Mac and Windows users. Podcasters can use it to normalize audio. This means they can make a quiet recording louder, or level two audio tracks recorded independently to the same volume. Audacity also allows podcasters to slowly fade music in and out of the recording. Because Audacity is popular, new podcasters can find online tutorials and downloadable plug-ins, which are software

HINDENBURG PRO

Hindenburg Pro is a podcast recording and editing program designed for professionals. It comes with a variety of features, including multiple audio tracks and transcripts. Podcasters must pay to use Hindenburg Pro. The cost of a subscription varies depending on which features the podcaster wants. Users describe the program as "intuitive" and nondestructive.[4] People who cannot figure something out can also find video tutorials to guide them through the process.

components that can be added to Audacity. These resources can help podcasters create the sound they want.

Another popular editing program is GarageBand, which is free and comes preinstalled on Apple products. It is easier to use than Audacity but doesn't provide as many advanced features. Someone looking for an extremely basic setup can use GarageBand on an iPhone.

A third program is Alitu, which can be used for both recording and editing. It is internet-based, so it can be used on any device. Alitu's programmers have focused on speeding up the editing process. Podcasters upload their intro and outro the first time they use the program. After that, Alitu adds it to each episode automatically. After a seven-day free trial, Alitu charges users a monthly fee.[5]

SHARING THE PODCAST

Once a podcast episode has been recorded, it must be uploaded to a hosting service. This is where the actual audio files are stored. Some podcasters may be tempted to simply upload the files to their own website. But the bandwidth demands of a popular podcast can be more than the host of a personal website can handle. Hosting services such as Blubrry or Libsyn will store podcasts long-term if podcasters pay a fee for an account. The host handles the RSS feed and distributes the podcast to various directories, such as

PODCAST WEBSITES

Listeners can often go to a podcast's website to learn more about the podcast and its host. Because the podcast website plays such an important role, it should be launched before the podcast is released. By putting the right information on the website, the podcaster shows that they're an expert on their podcast's topic, whether it's high school culture, eco-friendly products, or pet sitting. Once the podcast launches, the website acts as a central location for all podcast-related information, from show notes to episode lists. A website is key to building a strong online presence.

Apple Podcasts, Spotify, or Google Podcasts, so listeners can find it.

Podcasters must pay to use most hosting services. But some services, such as Buzzsprout, offer a free option. Before choosing a free plan, podcasters should read the service agreement carefully. Some free plans store individual episodes for only a limited time period before deleting them. Others insert ads into the podcast. The hosting service earns money from these ads, but the podcaster does not.

When filling out the forms for hosting services and directories, the podcaster must pick a category and possibly a subcategory. The podcast will be listed under this category in directories. For example, the category for a podcast might be science. Possible subcategories might include astronomy, mathematics, and social sciences. A carefully chosen category can help listeners find the podcast. Hosting services may also require show notes, which are displayed next to the episode description in apps and alongside the episode on websites. Show notes include a summary of the episode, information about the guests, and other helpful links to things mentioned throughout the podcast.

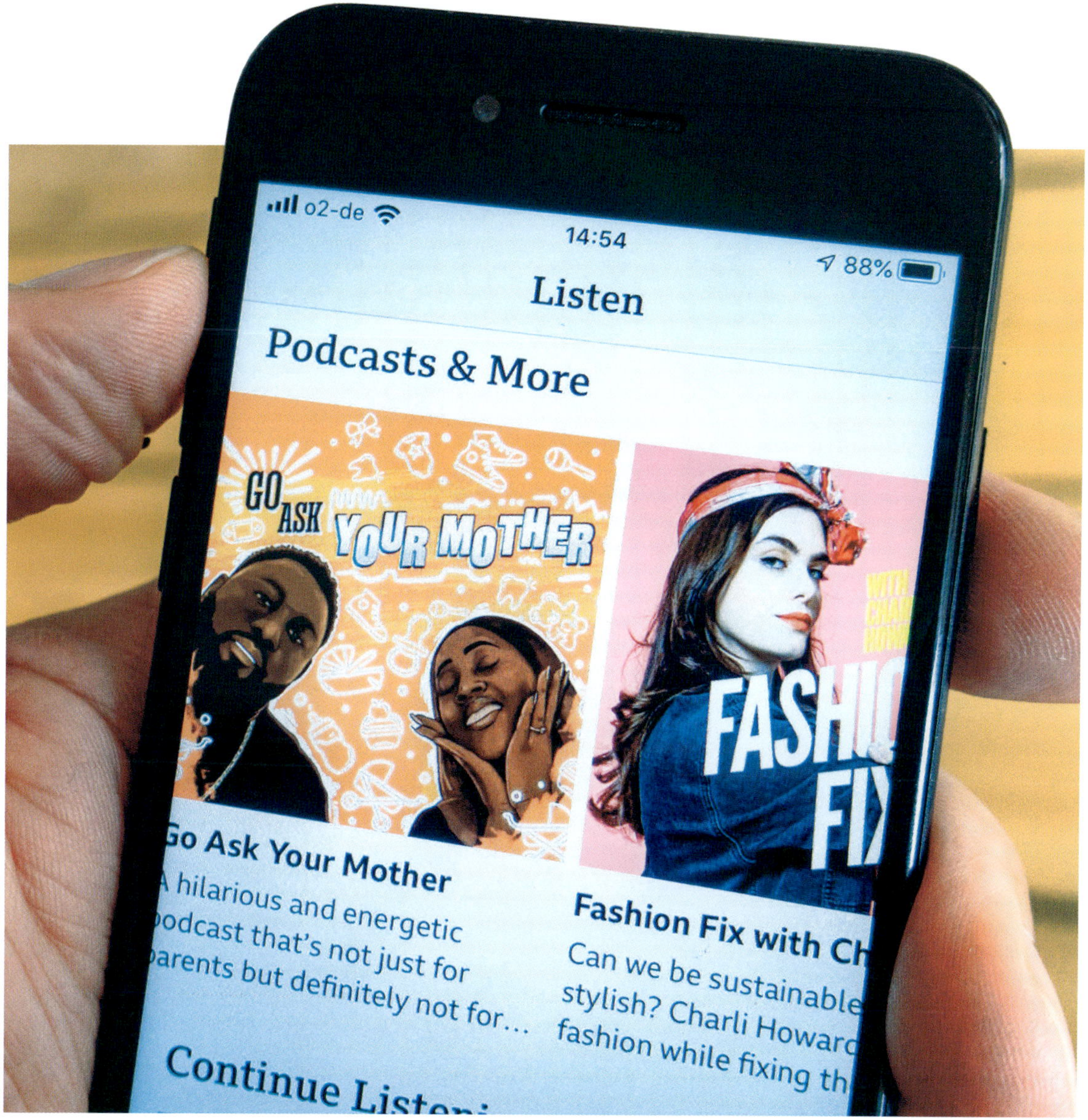

Podcasts can be found on many different directories, including BBC Sounds and iHeartRadio. These directories make it easy for users to sort, save, and listen to podcasts.

SXSW
MUSIC
FILM
INTERACTIVE

EARNING MONEY FROM PODCASTING

Many people wonder how much money a podcast can earn. At first, most podcasters make very little money. They may even lose money if they buy equipment such as microphones, pay for a hosting service, or hire someone to create their podcast artwork.

Certain well-established podcasts can make a lot of money. In 2018, *The Daily* podcast by the *New York Times* brought in $10 million in advertising.[1] The *New York Times* is an established newspaper that is a leader in print, digital, audio, and podcast reporting.

But podcasts don't need to be affiliated with major companies to turn a profit. In 2022, *The Joe Rogan Experience* by US comedian Joe Rogan was the number one podcast globally on Spotify.[2] Rogan is a controversial public figure for a variety of reasons. He has been criticized for making racist comments and speaking against COVID-19 vaccines. But people still tune into his podcast. Rogan has 200 million downloads a month and earns $30 million per year from his podcast alone.[3]

Some independent podcasters can expand their reach by attending events or recording live podcasts. Hrishikesh Hirway, *left*, hosts the podcast *Song Exploder*. He has interviewed musicians at events such as the SXSW Music, Film + Interactive Festival.

Hoping to earn an income from their podcasts, many new podcasters attempt to monetize immediately. They hear about other podcasters with big incomes and want to start earning money too. But Colin Gray from the website the Podcast Host advises that people wait until their podcasts are well established before trying to earn money. A podcast is established enough to monetize when it is listed in several major podcast directories, such as Spotify, Google Podcasts, and Apple Podcasts. The podcast should release a new episode every week and have at least 1,000 downloads per episode.[4]

But advertisers and sponsors want to see more than just downloads. They want to see engagement on social media platforms, which means they want to see listeners interacting with the podcast host and with each other. This interaction makes people feel involved. If listeners feel involved, they are more likely to support a podcaster financially. Once a podcaster has an established fan base and community, there are several ways to earn money on a podcast.

ADS, SPONSORSHIPS, AND AFFILIATE LINKS

An established podcaster can make money through advertisements and sponsorships. These advertisements are commercials intended to entice listeners to buy something. An ad might suggest that listeners buy a certain product, such as headphones for listening

Some podcasters advertise a company's product by featuring it in their shows. A podcaster who focuses on beauty might partner with a makeup company and review one of the company's products.

PODCAST ADS

Podcast ads are usually different from radio ads, which are often loud and lively. Radio ads are meant to grab the listener's attention. Podcast ads are meant to be part of a seamless podcast experience, since podcast listeners are already engaged. Spotify Advertising describes podcast ads as sounding like recommendations from a friend. The ad should end with a quick call to action, such as "Visit our website to find out more." The best podcast ads tend to be short, understated, and to the point.

to music. Other ads might suggest that listeners pay for a service, such as tutoring that specializes in ACT prep. The podcaster is paid to include the ad in their podcast.

Sometimes a podcaster will approach a specific business about advertising on the podcast. They present information about their podcast, including metrics about their listeners, and ask the company to pay for an ad. Podcasters can also work with a podcast advertising network, such as Authentic or AdvertiseCast. Some podcast hosting sites have their own advertising networks. The network finds advertisers, negotiates pay rates, and delivers ads to the podcaster.

Podcasters can accept several different kinds of paid advertising. Sometimes prerecorded ads are delivered to the podcaster. During the editing process, these ads are inserted into the podcast recording. The advertiser might request that the ad run at a specific place in the podcast, such as the beginning of the episode, the middle of the episode, or right before the outro.

If a company doesn't have a prerecorded ad, the podcaster may read a script that mentions the company's product or service. Sometimes there isn't a script, and the podcaster simply endorses the product in their own words. The podcaster might say, "When I record an upcoming episode, I won't use any headset except my Techlogic complete with microphone." Running an ad in a few episodes pays only a little money. Some podcasters try to find companies that will advertise on the podcast for a full season.

A sponsorship is different from an advertisement. The sponsor may pay money, but it may also donate goods or services. Sometimes these donations might be used to produce the podcast, especially if the sponsor donated equipment such as microphones or headsets. Other times, the donated items or services are used in giveaways. When a podcaster has a sponsor, the podcaster mentions the sponsor several times during each episode. The podcaster may say something such as, "Special thanks to our sponsor Carl's Computer Repairs, a local company that knows the difference between RAM and ROM." The owner of Carl's Computer Repairs knows that if someone listening to the podcast has a computer problem, there is a chance they will bring their computer to his business.

Podcasters can also use affiliate links to earn money. To do this, the podcaster joins a company's affiliate program. Then the company gives the podcaster special links for

> **"Sponsors pay based on the number of listeners that you have and where their ad is placed within the episode. Potential sponsors look at your per episode or per month downloads to get an idea of their listener base.[5]"**
>
> ***—Bridget O'Rourke, video producer for an online recording studio***

KICKSTARTER
What is Kickstarter?
Discover great projects
Start a project
Search projects
Bring creativity to life
Curious how Kickstarter works?
Learn more
Melissa's conceptual art project sent her on an Arctic expedition.
See all 329 Art proje
Staff Picks: Art
Painting an entire favela in Rio de Janeir
by Favela Painting in Rio de Janeiro, Brazil
With your help, we will paint an entire favela in R
de Janeiro!
Back to Rio

each product they mention in their podcast. When a podcaster who talks about hiking mentions their favorite brand of hiking boots, for example, they have an opportunity to post an affiliate link. The podcaster includes a link to the product in their podcast's show notes. When listeners click on the link, it takes them to the item. The link contains a special code so the podcaster gets credit when a listener uses the link and buys the product. The podcaster earns a percentage of each purchase.

The more listeners who click and make a purchase, the more income the podcaster earns. Amazon partnerships can yield an income as high as 10 percent.[6] These partnerships are also easy to use, since Amazon carries such a wide variety of products. The downside to affiliate programs is that the podcaster does not control them. If Amazon or another company with an affiliate program decides

"It's Amazon's way of saying thank you for bringing this customer to my website to make a purchase, here's some money for doing that.[7]"

—Whitney Bonds, blogger, on Amazon affiliate links, 2022

Crowdfunding on platforms such as Kickstarter can be a good way for podcasters to raise money for their shows. Some experienced podcasters even make podcasts about crowdfunding.

ADVERTISING A CROWDFUNDING CAMPAIGN

A podcaster who decides to raise funds on GoFundMe or Kickstarter must get the word out about the campaign in order for it to succeed. One of the best ways to do this is to volunteer as a guest on another podcast. This helps spread the word about the campaign to a new audience. To do this, the podcaster must find similar podcasts and reach out to the hosts. They should explain their campaign and ask the hosts if they can speak to their listeners.

to cancel the program or alter the terms, such as how long a link can be used, there is nothing the podcaster can do about it.

DONATIONS AND SUBSCRIPTIONS

Another way for podcasters to earn money is to go directly to their listeners. This is an option once podcasters gain a devoted following. Once this happens, the podcaster can ask listeners to make donations or become paid subscribers.

A podcaster can ask fans for donations through a crowdfunding campaign on GoFundMe, Patreon, or Kickstarter. This type of fundraising works well when a podcaster needs a large amount of money to finance a specific project, such as a special episode or upgraded equipment. It can also be used to fund the podcast throughout an upcoming year. The downside to crowdfunding is that it can require a lot of work, especially if the podcaster feels the need to create premium content to reward contributors. Another problem is that some crowdfunding campaigns simply never get off the ground.

Asking supporters for small donations is another way podcasters can make money. Some podcasters end each show by reminding listeners that their support is appreciated.

The most popular merch items for podcasts are T-shirts and other clothing items. These usually feature a podcast's artwork or logo. Podcasters may wear these merch items to promote their podcasts.

Listeners can make small regular donations through Patreon or Buy Me a Coffee. The podcaster can also include a PayPal or Stripe button on their website. Listeners can use these links to donate money to the podcaster through those payment services.

Another way to earn money is to charge a subscription for the podcast itself. This can be done on Apple Podcasts and Substack. Some podcasters offer both a free version of the podcast and a version for subscribers. The free version may include only part of a story or episode, while the remainder of the episode is available only in the subscriber feed. Other podcasters keep their main podcast free but offer premium content only to subscribers. The podcast *Comedy Bang! Bang!* does this by giving only subscribers access to some of their live shows. Podcasters may also offer ad-free episodes as an incentive to subscribe.

STUFF THAT SELLS

Some podcasters make money by selling merchandise, also known as merch. Among the most popular items are T-shirts, hoodies, mugs, phone cases, and stickers with podcast art printed on them. Other merch items might feature a phrase popularized on the podcast. The perfect merchandise for a podcast depends on what the podcast is all about. Fans of *Science Rules! with Bill Nye* can buy a stress ball shaped like a microscope. *LeVar Burton Reads* listeners can get busy writing their own stories with a *LeVar Burton*

Reads journal and pen. *Comedy Bang! Bang!* listeners can pick up a signed tour poster.

Some podcasters handle merchandise sales themselves, but this means they must order the items and ship out purchases. Other podcasters use a merchandise service that creates the items, takes orders, and ships out products. For example, the company Podswag brings in independent artists to help create podcast merchandise.

Podcasters can also make money by selling services. For example, a podcaster who likes teaching others about their topic and is good at explaining things might create an educational course. Another podcaster might prefer to act as a coach, talking one-on-one with a listener who is willing to pay for coaching. Podcasters can charge for organizing a workshop or for speaking at someone else's event. Podcasters who do this have usually developed a name and reputation and are selling their expertise.

Figuring out how to monetize a podcast can be as difficult as figuring out a podcast topic. What the podcaster chooses to do will depend on their skills, their interests, and the topic of their podcast. Not every type of monetization will work for every podcast or podcaster. That's why podcasters must carefully consider all the options.

KNOW YOUR AUDIENCE

A podcaster who wants to monetize must know who their audience is. After all, not everyone is interested in merch such as hoodies and mugs. Items with logos will sell only if they are the kinds of things that the audience uses. In the same way, services will sell only if they are something that listeners want and can afford. Smart podcasters keep their audience in mind not only when creating episodes but also when deciding how to monetize.

CHAPTER **EIGHT**

THE FUTURE OF PODCASTING

Today, podcasts come in all shapes and sizes. Many people create podcasts that are meant to be pure entertainment. But there are also other types of podcasts that offer content on every conceivable topic, from movies to true crime.

The good news is that the number of people listening to podcasts is growing. A 2021 eMarketer study released by Statista projected that 424 million people would be listening to podcasts every month by the following year. And by 2024, that number was expected to reach about 505 million.[1]

The overall makeup of podcast listeners is changing too. The company Edison Research gathers information that government departments and digital media companies use to make decisions. The group's 2021 Latino Podcast Listener Report found that 33 percent of US Latinos had listened to podcasts that were primarily in Spanish. That number rose from 24 percent in 2020.[2] These findings suggest that today's podcast listeners are seeking out and finding very specific content.

Experts say that as the number of podcasts continues to grow, many new listeners will be young people. Podcasters may be able to reach more of this age group by creating content on YouTube or social media.

The variety of what podcast listeners seek will also continue to grow. Gabriel Soto of Edison Research predicts that podcast listeners will become more varied. "These younger audiences, which are more ethnically diverse, will continue to transform the face of the podcast listener to reflect Black Americans, Latino Americans, Asian Americans, and other ethnic groups," Soto explained.[3]

Media experts are predicting changes beyond US audiences. They believe that the reach of many podcasts will become more global over time, increasing the demand for more multilingual content. Michelle Khouri is CEO and founder of FRQNCY Media, an audio production company that has worked with Coca-Cola, iHeart Media, and the Jane Goodall Institute. "I expect we'll see the globalization of podcasting really take hold next year," said Khouri in 2022. "I think we'll see podcasts succeeding in more languages than we've ever heard before and expressing the stories of vastly different cultures and geographic hubs. I'm especially excited about this prediction!"[4]

Media specialists also predict that future podcasts will become increasingly focused by topic. Kate Wojewoda-Celinska is the marketing manager for Spacelift, a company that manages cloud infrastructure.

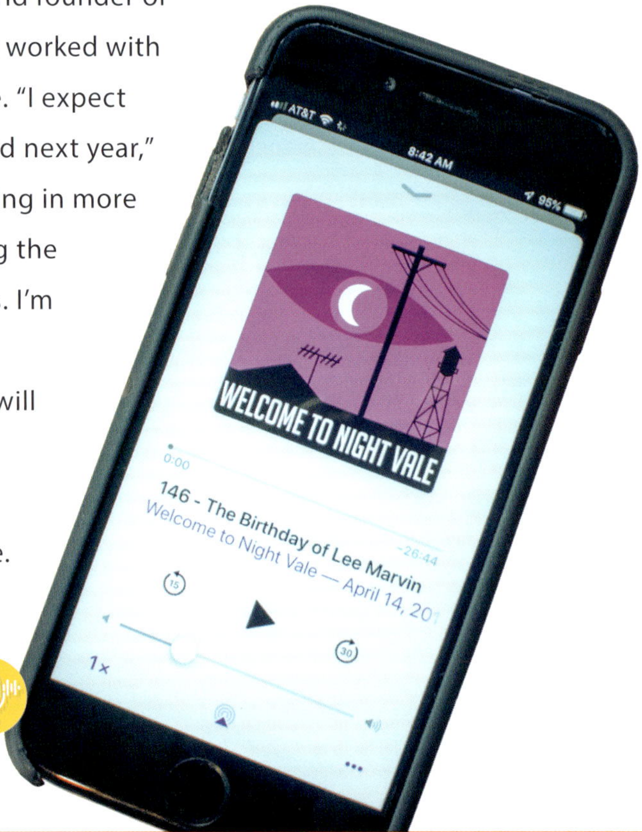

Future podcasters may choose to focus on very specific or unique topics within popular genres. For example, the fiction podcast *Welcome to Night Vale* tells the story of mysterious events in a small town.

"One trend that is likely to continue is the move toward niche podcasts," Wojewoda-Celinska said. "People are increasingly interested in tuning in to podcasts that cater to their specific interests, whether politics, cooking, gardening, or history. This means there will be a growing demand for unique and engaging content."[5]

DATA-DRIVEN DECISIONS

Experts expect to see more podcasters using listener preferences to shape their content. Yongming Song is CEO of the Chinese company Imgkits, which provides online photo editing services. "Content that resonates with the target audience is always high quality and desirable," he explained.[6] Song pointed out that collecting feedback from listeners can help podcasters create content that will encourage listeners to interact with the community.

To reach new listeners, podcasters may need to consider SEO, or search engine optimization, when creating scripts for their shows. Search engine optimization is a process used to push

LAW STUDENTS LEARNING

At the University of Virginia School of Law, Professor Michael Livermore runs the environmental podcast *Free Range*. Professor Kim Krawiec runs the podcast *Taboo Trades*, which is about controversial jobs. Both professors invite students to interview experts, conduct research, and help plan future episodes. This gives students the opportunity to earn class credit while taking a deep dive into difficult legal issues. They gain experience as podcast researchers and writers. "When you're writing a paper for a class, what you see about the issue is confined to sources you're able to find online. But when you get to talk to the actual players involved, you see the human side of the issue," student Matt Di Sandro said.[7]

A podcast episode description might include the episode title and number, a summary of the episode, and a list of guests featured on the episode. For a storytelling podcast, show notes might aim to create suspense or make listeners curious.

web content up the search results page. Search engines use programs known as bots to explore web pages and note what content those pages contain. The bots add this information to the indexes for each search engine. When someone types in a Google search for "podcast inventors," the engine reviews its indexes to give the person relevant search results. Each search engine uses a wide variety of indicators to move each site up or down the list of results. The goal of SEO is to move the site to the top of the page, where a person is more likely to click on it.

> **I think the money is going to be in podcast audience growth. . . . Any tools that can help boost podcast discoverability, especially on search engines, will be the future.[9]**
>
> ***—Fatima Zaidi, founder of a podcast production company, 2023***

Audio expert Rachel Corbett explains that putting the right keywords in a title helps a podcast episode appear in online search results. Podcast descriptions and transcripts should also be written with search keywords in mind. "Part of your keyword research should always involve typing that keyword into Google and seeing what pops up. If it's content similar to yours you know you've chosen the right one. If not, you might want to rethink it," wrote Corbett.[8]

This is why it is vital for podcasters to include an episode title and description on their podcast website. A transcript is even better because it gives the bots more content to explore. But the podcaster must also figure out which keywords people are using to find

Some podcasters who make video content create sets where they can record their conversations. They may need to purchase special video equipment, such as lighting, and do more editing at the postproduction stage. Other podcasters broadcast their videos live.

their podcast. When a listener looks for a science fiction podcast, are they searching for “science fiction” or “extraterrestrials”? Including both keywords in the episode description will let listeners find the podcast with either term.

There are many ways for podcasters to find popular search terms. One way is to check out the most popular podcasts and make a list of keywords found in their titles and descriptions. Another way is to use the automatic suggestions feature in Google

Podcasts and Apple Podcasts. All the podcaster needs to do is start entering a keyword and see what search term is suggested. An awareness of search terms is essential in an increasingly competitive podcast market.

MICRO-PODCASTS

Because there are so many hour-long podcasts, some creators have been moving toward shorter micro-podcasts. These are also called mini podcasts or short-form podcasts. They can be anywhere from one to 20 or 30 minutes long. The short podcast was made popular by the *New York Times* podcast *The Daily*. Short news podcasts make up only 1 percent of the podcasts recorded but 10 percent of what is downloaded.[11] The Reuters Institute at the University of Oxford specializes in journalism. It says that shorter podcasts engage younger listeners and increase listener loyalty.

REPURPOSING CONTENT

Another future podcasting trend will be repurposing content. This means podcasters will find ways to reuse their podcasts by creating other materials based on the same content. Repurposing content can help podcasters get the most out of their efforts and find new listeners.

The Podsicle Media Team says the easiest way to repurpose a podcast is to record a video while making the podcast. The podcaster can set up a webcam or phone to record the video while recording their podcast. The reason for doing this is simple. According to Search Engine Journal, the YouTube search function was the world's second-largest search engine in 2021.[10] A video recorded while making a podcast can easily be uploaded to YouTube, where potential viewers and listeners can find it.

Podcast content can also be used to create social media posts. This may include short audio or video clips. These clips could be made by editing existing video or audio files and selecting short, meaningful moments. Podcasters can also create graphics to post on social media platforms. These might include statistics pulled from an episode. For example, a podcast about applying to universities might include a statistic about the number of people who are accepted into their first-choice school. An image with the statistic might pull in listeners who want to learn more about the topic. Podcasters can also make graphics featuring quotes from the podcast. Any of these elements can be posted on social media to start conversations, build communities, and attract new listeners.

When considering how to repurpose content, podcasters shouldn't look at only the content they are currently creating. Motivate Talent is an agency that represents speakers and broadcasters. In 2022, it reported that 48 percent of monthly podcast downloads are older episodes.[12] Older episodes continue to be popular because new fans have not listened to them yet. That means these episodes can be rereleased or used to create audio clips, video clips, compilations, or graphics. Podcasters who learn

> **"Nerd out, know your show values, know your audience, and make something they can't get enough of (and can't get anywhere else!)[13]"**
>
> ***—Karen Burgess, member of a podcast production company, 2023***

to take full advantage of existing content can look at older episodes along with episodes currently in production.

Ambitious podcasters can also write blog posts or articles based on their podcast topics. The more ways a creative podcaster can reuse content, the more opportunities they will have to engage possible listeners. The point is to get as much use out of existing content as possible.

ADS MATTER

Not all projected changes in the world of podcasting are positive. By the end of 2022, people were spending a lot of money on advertising in various types of media. But turmoil affects economics, and economics affects advertising. This turmoil could be war, such as the war in Ukraine. This conflict started in 2014 and escalated in February 2022, when Russia invaded Ukraine. The war led to higher energy prices. Higher energy prices led to inflation, which meant that everything from vegetables to houses cost more.

It isn't just individuals who must spend more to get the things they want. Businesses are affected too. This means businesses will likely look for ways to save money. For some, this means spending less on advertising, including podcast advertising.

This doesn't mean that there will be no advertising. An eMarketer study estimated that in 2023, podcast ad revenue, or income, would grow by almost 29 percent.

That would be the same as the growth rate in 2022 but only half the growth rate in 2021.[14] This suggested there was still money to be made in podcasting. But the income was leveling off and expected to drop in 2024. For people expecting to make big money in podcasting, this wasn't good news.

Nevertheless, there is still room in podcasting for creative individuals who are passionate about topics they want to share. These people may be able to make some money by podcasting. If they create quality podcasts and put them out into the world, they can reach a growing community of listeners. These listeners may be in their own neighborhoods or on the other side of the world. Either way, podcasting is a solid way to develop communities based on common interests. It is an example of global media, connecting people across places and spaces.

BIG INFLUENCE

Many social media influencers have created their own podcasts. Examples include *Charli and Dixie: 2 Chix* with influencers Charli and Dixie D'Amelio, and *Capital University* with TikTok star Bryce Hall. Media experts expect more social media influencers to start podcasting. This is because influencers know how to use social media to gain more followers and monetize their content. With a growing number of listeners, podcasts are a good way to do this.

ESSENTIAL **FACTS**

MAKING PODCASTS PROFESSIONALLY

- A professional organization or business develops an idea for a podcast that will help it expand its reach, market its products, or spread its message.
- The business or organization assembles a team to record and produce the podcast. This may include producers, writers, hosts, audio engineers, and graphic designers.
- The podcast team works to figure out the project's budget, scripts, episodes, schedule, guests, and other logistics.
- The team records the podcast. To get the best sound quality, professionals often record in a soundproofed studio.
- The podcast enters the postproduction stage. An audio engineer edits the recording and adds music and special effects. The finished podcast is uploaded to a hosting platform and released to listeners.

MAKING PODCASTS INDEPENDENTLY

- An individual develops an idea for a podcast. The podcaster must choose a topic they are interested in and consider how broad or narrow the focus of the podcast should be.
- The podcaster brainstorms episode ideas, chooses a podcast format and title, and sets a publishing schedule. The podcaster chooses which equipment to use and sets up a recording studio.

- The podcaster writes a script, creates the intro and outro, and makes sample recordings. Then they record the podcast.
- The podcaster edits the recording and adds sound effects or music, often using a computer program such as Audacity or GarageBand.
- The podcaster uploads the completed podcast to a hosting service to be shared with listeners. The podcaster may try to monetize their podcast through ads, subscriptions, merch, or sponsorships.

QUOTE

"People are increasingly interested in tuning in to podcasts that cater to their specific interests, whether politics, cooking, gardening, or history. This means there will be a growing demand for unique and engaging content."

—Kate Wojewoda-Celinska, Spacelift marketing manager, 2023

GLOSSARY

affiliate
Connected to or controlled by a group or business. Often used to describe a business arrangement in which a creator gets a bonus for every sale they generate for a business.

cassette
A plastic case that holds spools of audio tape or video tape.

crowdfund
To fund a project by raising money through donations or purchases from a large number of people.

directory
A list or guide of all items available in a certain platform or place.

ducking
In audio, when music fades into the background when someone begins to speak.

endorse
To publicly support or approve of something.

frequency
The pitch of a sound, or the number of times a sound wave repeats.

globalization
The processes by which trade and technology make the world and its people more interconnected.

metrics
Data used to measure the performance of a business, organization, individual, or project.

monetize
To use something to make money or to earn money from something.

patent
The exclusive right to an invention.

postproduction
Work done on a piece of media after filming or recording has taken place.

saturated
Describing a market that is full.

serialized
Broadcast in regular installments or segments.

splice
To insert one audio recording into another.

sync
A shorthand term for synchronize, or to enable devices to work together.

transcription
The conversion of spoken words into written words.

ADDITIONAL **RESOURCES**

SELECTED BIBLIOGRAPHY

Foster, Katelyn. "Podcasts Offer New Brand of Interactive Learning for Students." *University of Virginia School of Law*, 7 Apr. 2023, law.virginia.edu. Accessed 5 Aug. 2023.

Morris, Tee, and Chuck Tomasi. *Podcasting for Dummies*. John Wiley and Sons, 2017.

Weldon, Glen. *NPR's Podcast Start Up Guide*. Ten Speed Press, 2021.

FURTHER READINGS

Conley, Kate. *Social Media and Modern Society*. Abdo, 2022.

Currie, Stephen. *Digital Literacy: What Is It and Why Does It Matter?* ReferencePoint, 2024.

Jaskulka, Marie. *Making Web Content*. Abdo, 2025.

ONLINE RESOURCES

Booklinks
NONFICTION NETWORK
FREE! ONLINE NONFICTION RESOURCES

To learn more about making podcasts, please visit **abdobooklinks.com** or scan this QR code. These links are routinely monitored and updated to provide the most current information available.

MORE INFORMATION

For more information on this subject, contact or visit the following organizations:

The Berlin School of Podcasting
33 Körtestrasse
Berlin-Kreuzberg, Germany 10967
berlin-school-of-podcasting.com
The Berlin School of Podcasting teaches audio storytelling as an art, offering online classes and a four-week intensive summer school.

Syracuse University
900 South Crouse Ave.
Syracuse, NY 13244
syracuse.edu
Syracuse University offers degrees in broadcast and media journalism, preparing graduates for work in radio, television, and online platforms.

Toastmasters International
9127 S. Jamaica St., Ste. 400
Englewood, CO 80112
toastmasters.org
Toastmasters International is a nonprofit educational organization that helps people perfect the skills they need to be top-notch public speakers and leaders.

SOURCE **NOTES**

CHAPTER 1. PODCAST PROJECT

1. "Audacity Tutorial For Beginners." *YouTube*, uploaded by Mike Russell, 21 Jan. 2022, youtube.com.

CHAPTER 2. THE HISTORY OF PODCASTS

1. "Sony Walkman." *Design Museum*, n.d., designmuseum.org. Accessed 1 Nov. 2023.
2. Eric Griffith. "A Visual History of the Apple iPod." *PC Magazine*, 10 May 2022, pcmag.com. Accessed 1 Nov. 2023.
3. Annalee Newitz. "Adam Curry Wants to Make You an iPod Radio Star." *Wired*, 1 Mar. 2005, wired.com. Accessed 1 Nov. 2023.
4. "About: Daily Source Code." *DBpedia*, n.d., dbpedia.org. Accessed 1 Nov. 2023.
5. "Apple Takes Podcasting Mainstream." *Apple*, 28 June 2005, apple.com. Accessed 1 Nov. 2023.
6. "Apple Takes Podcasting Mainstream."
7. Paul Boutin. "Steve Jobs Keynote Live from MacWorld 2006." *Engadget*, 19 July 2019, engadget.com. Accessed 1 Nov. 2023.
8. Boutin, "Steve Jobs Keynote."
9. "Podcasting Historical Timeline and Milestones." *International Podcast Day*, n.d., internationalpodcastday.com. Accessed 1 Nov. 2023.
10. Katrina Bishop. "How 'Serial' Is Making Podcasts Mainstream." *CNBC*, 21 Nov. 2014, cnbc.com. Accessed 1 Nov. 2023.
11. Lex Friedman. "Apple: One Billion iTunes Podcast Subscriptions and Counting." *Macworld*, 22 July 2013, macworld.com. Accessed 1 Nov. 2023.
12. Colleen Walsh. "The Podcast Revolution." *Harvard Gazette*, 27 Oct. 2011, news.harvard.edu. Accessed 1 Nov. 2023.

CHAPTER 3. DEVELOPING A PROFESSIONAL PODCAST

1. Lex Friedman. "Apple: One Billion iTunes Podcast Subscriptions and Counting." *Macworld*, 22 July 2013, macworld.com. Accessed 1 Nov. 2023.
2. Tony Cohn. "Making the Smithsonian's New 'Sidedoor' Podcast Series a Reality." *Smithsonian Insider*, 26 Oct. 2016, insider.si.edu. Accessed 1 Nov. 2023.
3. "Podcasts." *Smithsonian*, n.d., si.edu. Accessed 1 Nov. 2023.
4. "Podcasts." *Library of Congress*, n.d., loc.gov. Accessed 1 Nov. 2023.
5. Alison Osborne. "8 of the Best Branded Podcasts and Why They're Successful." *CoHost*, 22 Sept. 2023, cohostpodcasting.com. Accessed 1 Nov. 2023.
6. "Why Your Business Needs a Podcast: Podcasting for Business 2019." *YouTube*, uploaded by Pod Sound School, 22 May 2019, youtube.com.

CHAPTER 4. RECORDING A PROFESSIONAL PODCAST

1. "#NCW2023 Interview: What It's Like Being a Podcast Producer." *YouTube*, uploaded by ERIC, 15 May 2023, youtube.com.
2. Abel Grunfeld. "Podcast Advertising: The What, Why and How of Podcast Ads." *Riverside*, 17 Oct. 2023, riverside.fm. Accessed 1 Nov. 2023.
3. "7 Tips For Conducting Better Podcast Interviews." *YouTube*, uploaded by Buzzsprout, 22 Mar. 2021, youtube.com.

CHAPTER 5. DEVELOPING AN INDEPENDENT PODCAST

1. "Popular and Trending." *Google Podcasts*, n.d., podcasts.google.com. Accessed 1 Nov. 2023.
2. Alban Brooke. "How to Start a Podcast: Complete Step-by-Step Guide [2023]." *Buzzsprout*, 3 July 2023, buzzsprout.com. Accessed 1 Nov. 2023.
3. "How Long Should Your Podcast Episode Be?" *YouTube*, uploaded by Think Media, 28 Jan. 2022, youtube.com.
4. Brooke, "How to Start a Podcast."
5. Brooke, "How to Start a Podcast."
6. "Podcast Names: Picking the Right Name to Grow an Audience." *YouTube*, uploaded by The Podcast Host, 19 Oct. 2020, youtube.com.

CHAPTER 6. RECORDING AN INDEPENDENT PODCAST

1. "How to Start a Podcast for FREE (Using Your Phone)." *YouTube*, uploaded by Think Media, 9 Aug. 2021, youtube.com.

2. "How to Record a Podcast for Beginners: Ultimate Podcast Guide for Beginners." *YouTube*, uploaded by Katie Steckly, 4 Aug. 2020, youtube.com.

3. "10 Things I Wish I Knew Before I Started a Podcast." *YouTube*, uploaded by Buzzsprout, 13 Aug. 2020, youtube.com.

4. "Best Podcast Recording & Editing Software." *YouTube*, uploaded by Buzzsprout, 13 July 2020, youtube.com.

5. "Everything You Need to Podcast FREE for 7 Days." *Alitu*, n.d., alitu.com. Accessed 2 Nov. 2023.

CHAPTER 7. EARNING MONEY FROM PODCASTING

1. Brett Robinson. "How Much Do Podcasters Make?" *Wired Clip*, 11 Sept. 2023, wiredclip.com. Accessed 2 Nov. 2023.

2. Marisa Dellatto. "'The Joe Rogan Experience' Was Spotify's Top Podcast—Despite Protests about Him Being on the Platform." *Forbes*, 30 Nov. 2022, forbes.com. Accessed 2 Nov. 2023.

3. Robinson, "How Much Do Podcasters Make?"

4. Robinson, "How Much Do Podcasters Make?"

5. "Podcast Sponsorship (Everything You Need to Know)." *YouTube*, uploaded by Riversidefm, 5 Sept. 2022, youtube.com.

6. Emily Schwake. "9 Strategies for Monetizing a Podcast, from Sponsorships to Premium Content." *Wix Blog*, 4 Jan. 2023, wix.com/blog. Accessed 2 Nov. 2023.

7. "Watch This BEFORE You Sign Up to the Amazon Affiliate Program: Amazon Affiliate Tutorial 2023." *YouTube*, uploaded by Whitney Bonds, 7 Nov. 2022, youtube.com.

CHAPTER 8. THE FUTURE OF PODCASTING

1. Tom Webster. "Pods Heard round the World." *Sounds Profitable*, 10 July 2022, soundsprofitable.com. Accessed 2 Nov. 2023.

2. Webster, "Pods Heard round the World."

3. Marina Hanna. "2023 Podcast Industry Forecast." *Pacific Content*, 21 Dec. 2022, pacific-content.com. Accessed 2 Nov. 2023.

4. Hanna, "2023 Podcast Industry Forecast."

5. Stewart Townsend. "What Is the Future of Podcasting in 2023." *Podcast Hawk*, 12 Dec. 2022, podcasthawk.com. Accessed 2 Nov. 2023.

6. Townsend, "What Is the Future of Podcasting in 2023."

7. Katelyn Foster. "Podcasts Offer New Brand of Interactive Learning for Students." *University of Virginia School of Law*, 7 Apr. 2023, law.virginia.edu. Accessed 2 Nov. 2023.

8. Rachel Corbett. "A Beginner's Guide to Podcast SEO." *Rachel Corbett*, n.d., rachelcorbett.com.au. Accessed 2 Nov. 2023.

9. Hanna, "2023 Podcast Industry Forecast."

10. Dave Davies. "Meet the 7 Most Popular Search Engines in the World." *Search Engine Journal*, 3 Mar. 2021, searchenginejournal.com. Accessed 2 Nov. 2023.

11. Katie Paterson. "Mini-Podcasting Is on the Rise—Here's What You Need to Know." *Podcast Host*, 29 July 2022, thepodcasthost.com. Accessed 2 Nov. 2023.

12. "The Future of Podcasting: 7 Trends to Watch." *Motivate Talent*, n.d., motivatetalent.com. Accessed 2 Nov. 2023.

13. Hanna, "2023 Podcast Industry Forecast."

14. Ariel Shapiro. "Podcasting Could Be in for a Rocky 2023." *Verge*, 26 Dec. 2022, theverge.com. Accessed 2 Nov. 2023.

INDEX

ABOUT THE **AUTHOR**

SUE BRADFORD EDWARDS

Sue Bradford Edwards is a nonfiction author who writes about culture, history, and science. She is the author of 25 other titles from Abdo Publishing, including *Ancient Maya*, *Hidden Human Computers*, *Cancel Culture*, *The Evolution of Reptiles*, and *The Evolution of Mammals*. Her favorite podcasts tend to group around science, history, and fiber. She listens to the podcasts *Stuff to Blow Your Mind, The Past and the Curious,* and *Fiber Nation*.